LIGHTNINGS LIVE ON!

Compiled and Edited by

HUGH TREVOR

for the

LIGHTNING PRESERVATION GROUP BRUNTINGTHORPE

This book is dedicated to Suzanne and to my Mother
for putting up with it all with (fairly) good grace

First published 1996

ISBN 0 9528829 0 6

Published by the Lightning Preservation Group, Bruntingthorpe

Contact telephone: 01243 827266

***Cover photographs**:*

***Front**: Group Captain John Spencer banks Lightning F.6 XR728 over the cloud-covered Lincolnshire Wolds for the camera of Flt Lt Ian Black in a T.5.* [Ian Black]

***Rear, Top**: The final Lightning 3-Ship formation from British Aerospace Warton on 16th December 1992; XS904 is nearest the camera, XP693 is leading the formation and XR773 is on the right.* [Ian Black]

***Rear, Below**: 74 Squadron Lightning F.1 XM145 'Q' on reheat test at the then English Electric Aviation Co. airfield at Warton in 1960.* [British Aerospace/North West Heritage Group]

The Firebirds aerobatic team of 56 Squadron brought Lightnings and the thunder of jet noise before an appreciative public in 1963—at which time it appears the poor groundcrew were not issued with ear defenders! In the upper photograph, Lightnings of 111 Squadron are also visible in the distance.
[Top: British Aerospace; Bottom: Rolls Royce plc]

CONTENTS

ACKNOWLEDGEMENTS

Without the support of the Walton family at Bruntingthorpe Airfield, the Lightning Preservation Group as such would probably not exist. Our grateful thanks go to them for all their help over the last eight years and hopefully into the future. Their continuing support of Nick Grace's vision of a working aviation museum at Bruntingthorpe under the banner of British Aviation Heritage has enabled a unique collection of classic British ex-military aircraft to be assembled, owned by a variety of groups of people but with the common aim of keeping their own particular aircraft as far as possible in a serviceable condition. Without the Walton family, some of these irreplaceable aircraft could have been reincarnated as beer-cans. Nev Martin's Phoenix Aviation is also based at Bruntingthorpe and Nev's team has given the LPG much practical help over the years.

I should also like to express my gratitude to those who contributed articles, as otherwise there would be no book. It would be iniquitous to single out individuals, as so many have provided enormous help well above and beyond the call of duty; when all is said and done, they did it because they all love the Lightning. Much help was also received from Lightning historians and enthusiasts, in particular Stewart Scott, John Jackson and Terry Senior. Sincere thanks to Pam Kaye and John & Lin Booth for help with word-processing and to Richard Crossley for many hours spent in the darkroom. Graham Jackson gave invaluable help over many hours with desk-top publishing and Geri Jackson refreshed the inner men at frequent intervals. Advice was also generously provided by Richard Clarkson of the Vulcan Restoration Trust at Southend.

Grateful thanks to Sqn Ldr Peter Singleton & Don Blevin of the Air Historical Branch (RAF), Andrew Siddons of Rolls Royce Public Affairs (Derby) and Flt Lt Russ Parker of DTEO Shoeburyness for their patience and hospitality; also to Mr Bill Hunt of CS Photography, ex-Firebird Brian Cheater, and Flt Lt Mark Simmonds for his help when History Officer of 5 (Fighter) Squadron. Ken Ellis generously supplied updated information on Lightning relics.

Current and former employees of British Aerospace have been very helpful with respect to the book, in particular Ian Lawrenson of the British .Aerospace North West Heritage Group at Warton & Samlesbury; Colin Charnley, Wally Rouse, Joe McGrath, Andy Waite, Chris Foster, Jim Dickinson, Mick Vincent, Brian Lewis, Jim Britton, Alan Sands, Alan Hutchinson and last, but definitely not least, Bas Livesey, all of whom made us feel very welcome and provided much information during base visits.

Finally, no book on the Lightning would be complete without a contribution from Wg Cdr 'Bee' Beamont, who very kindly and speedily agreed to write the Foreword; both 'Bee' and Jimmy Dell also gave advice and drew my attention to inaccuracies in early drafts, for which I am very grateful. Any errors which remain are entirely of my own making.

PHOTOGRAPHS

Photographs credited 'Crown Copyright/MOD' are reproduced with the permission of the Controller of HMSO. Thanks also to British Aerospace plc, Rolls Royce plc, Blackpool Gazette & Herald Ltd, Lancashire Evening Post Ltd, and other organisations and individuals who gave permission for their photographs to be reproduced.

Lightning photographers and enthusiasts were generous with the loan of unique photographs and particular mention must be made of Terry Senior, Stewart Scott, John Jackson, Karl Kaehler, Oliver Zarden, Dave Drummond, Trevor Garrod, Steve Gensler, Peter Talbot, Geoff Hill and Peter & Margaret Thompson. Flt Lt Ian Black generously provided several photographs including the superb shots which grace the cover, while Falcon Aviation came up trumps with regard to Saudi aircraft. Particular thanks are due to MAP and APN: Brian Pickering of Military Aircraft Photographs (Westfield Lodge, Aveland Way, Aslackby, nr Sleaford, Lincs. NG34 0HG; Tel. 01778 440760) has provided me and countless enthusiasts with hundreds of Lightning photographs over the years, a current listing of which can be found in the MAP British Military Catalogue; Brian Stainer of Aviation Photo News (56 Tachbrook Road, Feltham, Middx. TW14 9PB; Tel. 0181 890 1880) lists 325 Lightning B/W photographs in Catalogue M.1 'EE Lightning'. Both MAP and APN are highly recommended for those chasing the 'Holy Grail': a photograph of every Lightning built!

Great care has been taken to credit the source of photographs where known, however over the years many photographs have been acquired which are of unknown origin. If one of yours is published here without acknowledgement, please accept my sincere apologies.

FOREWORD

Wing Commander ROLAND BEAMONT

CBE DSO* DFC* DL FRAeS

Formerly Chief Test Pilot of English Electric Aviation
and later Director of Flight Operations for BAC and Panavia

The cockpit of Lightning Mk.6 XR728 was comfortingly familiar. With seat height adjusted for good forward view, the comfort level was relaxed with all controls and switches within easy reach. The Avons hummed smoothly at idle/fast idle and the engine panel showed all normal readings. With cabin auto-temperature controller set in the normal range the airflow was fresh, comfortable and uncontaminated by fuel or other odours.

Taxying out with the sadly now outmoded stick-lever brakes had been a smooth and undemanding operation, and now the Lightning was on the runway 'numbers' and ready to go.

Throttles up to max. dry and the Avons' thunder replaces the 'idling' hum. JPT's checked and within limits. Final checks of the engine panel, cabin pressure, oxygen and flaps. Air brakes closed. Release the brakes. Light the blue touch-paper (alright—engage reheat). Down the centre-line straight as a die; 0-140 knots in 10 seconds—rather better than the traffic-light Grand Prix—then ease the stick back at V_r. Nosewheel off, and then quickly put it down again!

This was not one of a thousand exhilarating Lightning take-offs on test from Warton, but a 'taxi-run' at Bruntingthorpe nearly 20 years after my last Lightning test flight. The temptation to continue the rotation into a smooth lift-off, clean up undercarriage and then flaps quickly to avoid embarrassment with the flap blow-up valve at 250 kts, and then rocket to the tropopause was (almost) as great as any temptation in life.

But Bruntingthorpe's three thousand yards felt more like three hundred yards in a rejected Lightning take-off; so out of reheat to idle, pull brake 'chute handle and—thank God for that as the 'chute develops, giving a strong impression of running into a haystack. Then wheel-brakes; how much easier than the now universal American-style individual rudder-pedal brake motors. And so to jettison the 'chute, and later have a chap say: "Why couldn't you delay the drag 'chute so that I could get it on camera?" You couldn't please everyone at a Farnborough Display either!

It had been sheer nostalgia all the way—except for the last part; I'd only aborted take-off once at Warton in 14 years of Lightning testing, when a canopy self-jettisoned on the T.4 prototype.

It was a tremendous privilege and pleasure to drive the 'Bruntingthorpe Lightning' and to experience the enormous enthusiasm of the Bruntingthorpe Group and their skilled professionalism. I wish them all continued success and good luck for the future.

Has any other jet aircraft in history ever generated so much enthusiasm and devotion?

'Bee' Beamont & Andy Williams share a joke with Ray Nightingale of the LPG at Bruntingthorpe, 24th June 1990. [HT]

Roland Beamont

February 1996

INTRODUCTION

THE LIGHTNING PRESERVATION GROUP

The origins of the Lightning Preservation Group date to the summer of 1987 when several enthusiasts who were gathered at the Binbrook crash gates began to contemplate a bleak future of no more Binbrook, no more Lightnings. The only way to compensate for this impending loss seemed to be to buy one! The idea grew, and more potential members were recruited (as sources of finance!) as the focus changed from the purchase of a spares-recovered F.3 from the decoy line to delivery of a complete and fully-functional F.6.

The dream was consummated on 24th June 1988 with the delivery to Bruntingthorpe of XR728, favourite mount of RAF Binbrook Station Commander Group Captain John Spencer, complete with "personalised" JS tailcode. Originally the aircraft was to be maintained in static condition as the Group did not have the finance or the expertise to maintain a flying aircraft, even if the CAA had been better disposed towards Lightnings flying on the civil register. However, the ex-Binbrook personnel in the Group suggested that the best way to preserve the aircraft, especially if permanent hangarage were not available, was to run the engines to keep the systems operating and to drive out moisture. The Group then progressed to taxying 728, and eventually to the use of reheat and performing fast taxi-runs along Bruntingthorpe's 10,000ft runway in front of the public at biannual open days.

XR728 is slowed by the brake parachute on Bruntingthorpe's 10,000 ft runway. The 'chute takes less than 5 seconds to deploy and 2 hours to pack ! [HT]

Late in 1992 it became apparent that the four remaining Lightnings flying at BAe Warton would shortly be up for tender and that another phase of Lightning operations would be coming to an end. Similar emotions to those experienced at the closure of Binbrook were in evidence in Group members and there was a heated discussion as to whether we should tender for a second Lightning, or whether this would leave us overstretched financially and with regard to maintenance effort, and would money not be better spent trying to put a roof over XR728? A compromise was agreed whereby the Group would put in a relatively low bid for XS904, a high-fatigue airframe which would not be required as a potential 'flyer' by Barry Pover's team which would bid for the other three aircraft. The bid for 904 was seen more as an insurance against it ending up in the hands of a scrapman—or so we told ourselves. In reality we became desperately keen to get the aircraft and were ecstatic on 21st January 1993 when XS904 swept low across Bruntingthorpe Airfield on its delivery flight, accompanied by Group Captain Andy Williams in a Tornado F.3.

The Group's third major acquisition occurred a year later when the QRA hangar from the former Lightning base at RAF Wattisham was obtained, thanks to the generosity of Trafalgar House Construction and with the assistance of Gale Construction Ltd. Our current challenge is to re-erect this ideal hangarage for our Cold War Warriors, and the aim of this book is to assist the fund-raising effort for this and for continued maintenance.

XR728 on reheat tests, 24th March 1991 [Tony Hulls]

The Lightning has always been a Man's Aircraft to maintain and has an amazing ability to draw blood on the most minor task. As our two aircraft age gracefully they are tending to require more effort and expertise and the Group is always interested to hear from engineers and electricians, Lightning or not, preferably living within commuting distance of Bruntingthorpe and with the time, knowledge, enthusiasm and commitment to help maintain the aircraft in serviceable condition into the future.

Various members of the LPG pose in front of XS904:
(L-R) Hugh Trevor, Chris Norris, Max Waldron, Jack Read, Ricky Wolfe, Chris Summers, Andy Bonsell, Richard Norris, Jim Featherby and Bob Tuck [HT]

This book outlines the development of the Lightning and includes articles and anecdotes relating to the operational and trials use of this outstanding British fighter, together with a review of Lightning preservation following its withdrawal from service.

Lightnings live on: in stories and photographs, in our memories—and at Bruntingthorpe.

OPERATION CORRUGATE—
QUEST FOR A 'Q' SHED

RICHARD NORRIS
President of the Lightning Preservation Group

I think most Lightning addicts would agree that two Mk.6's on live QRA was the ultimate spectacle from the enthusiasts' viewpoint. The Q-Shed was always a special place to visit when "live" and one could sense the adrenaline in the atmosphere. I assume that for many pilots and groundcrew a 'Q' launch was a special experience, the culmination of many years of training to perfect a quick and efficient live interception, although the realist in me knows that the hours of waiting must have been exceptionally boring; made worse for the pilots in that too many mugs of tea could not be downed due to the unknown flight time, coupled with the lack of an in-flight portaloo (although I understand that emergency receptacles were carried). My all-too-rare visits to the Binbrook Q-Shed certainly left a deep impression on me and from the inception of the Lightning Preservation Group (LPG) in the summer of 1987, the idea of preserving a Lightning in a Q-Shed was brought up in conversation from time-to-time (usually by me). Other members would politely discuss this 'ultimate museum' but we would quickly dismiss the idea as totally unrealistic.

A wonderfully atmospheric shot of two 5 Squadron pilots preparing for a QRA scramble on a wet night at Binbrook [Crown Copyright/MOD]

The demise of front-line Lightning operations at Binbrook in April 1988 resulted in the journey from Bedfordshire to Lincolnshire becoming just a happy memory, as was the view from Crash Gate 2 of the green corrugate-clad building. Some four years later, on Saturday 4th July 1992, I visited RAF Wattisham as XP693 was to make her last-ever visit for a 56 Squadron Families day, prior to 56's disbandment and the RAF relinquishing the station. Departure of 693 to Warton was on the Monday, which was a warm summer's day with a near clear blue sky. Peter Gordon-Johnson was the pilot and the departure was memorable.

Peter Gordon-Johnson performs a pre-flight inspection of XP693 at Wattisham on 6th July 1992 [Richard Norris]

XP693 was first to take-off, followed by a 56 Squadron Phantom and a Tornado F3; the three-ship formation overflew the airfield twice before departing for some air to air photography.

Peter Gordon-Johnson in XP693 leads a Phantom FGR2 and Tornado F3 in a photo sortie to mark 56(F) Squadron relinquishing both the Phantom and RAF Wattisham. [Geoff Lee/BAe]

Again, whilst this event in itself was totally absorbing, I could not help but wonder what would be the fate of Wattisham's Q-Shed after the RAF had vacated the airfield? Mick Cameron, who was then based at Wattisham, told me that after the RAF's departure the Army Air Corps would be moving to the airfield and he added that it looked unlikely that the Army would have a use for the building. Very interesting......an initial approach was made to Captain (now Major) J Ward, Quartermaster of 3 Regiment, AAC, who was very sympathetic when I explained that the LPG wished to acquire the Q-Shed. He explained that it was planned to demolish the building and replace it with a Search and Rescue Hangar. The first complication, however, was that although the Army was responsible for the airfield the demolition and construction of the new building was still to be the responsibility of the RAF. Despite this, it was suggested that I might wish to visit Wattisham in order to have a detailed look at the building.

I must admit I thought our enquiry was a bit of a long shot, but determination had to be the order of the day. The building appeared to be in remarkably good condition and the main doors had been replaced quite recently. Our initial approach to the Army was at the end of July, and it was the second week of September before an official reply to our enquiry was received from Strike Command. Oh dear! Or words to that effect......we were informed that the demolition and construction programme was too far advanced to amend the plans. Whilst I totally understood the RAF's position it was one that I could not accept if the building were to be saved.

A certain Lightning F.Mk.6 resides at Bruntingthorpe Airfield tailcoded 'JS'. Whilst the aircraft had held up well since its arrival in June 1988 signs of deterioration were beginning to show in its reliability, especially with engine starting. A roof over its head would help remedy the situation. Previous 'owner' Air Commodore John Spencer was unable to "pull strings", but would ensure that when the tender was awarded for demolition of the Q-Shed that we should be able to contact the successful tenderer and negotiate prices and terms direct. At the end of November we were informed that Trafalgar House Construction would be handling Q-Shed demolition and construction of the Search and Rescue Hangar, and so we presented our objective of saving the building to the company. They listened with interest, informing us that our proposal would be discussed at their next board meeting and that an answer to our request and a quotation would be forthcoming by the middle of December.

Deep joy! Not only were Trafalgar House prepared to donate the building to the LPG, as they understood its historic importance, but their sub-contractors Gale Construction would dismantle the building at no charge to the Group! Pure elation, Cloud Nine, etc., etc. The Group would, however, have to cover a percentage of the loading costs and the total carriage costs from Wattisham to Bruntingthorpe which, when all is said and done, was the very least we could expect. We felt, and still feel, very privileged to have received such generous treatment.

Question: How many trailers do you think we'll need to move the building? Answer: One bloody big 'un about 110 x 70ft. This was the type of question I was asking Trafalgar House and Gale Construction and, seriously, they estimated about 5 trailers should do the job. In the event we needed one more to take the large beams from which the main doors hang, each measuring some 55ft and requiring an 'extender trailer'. You live and learn. At one stage it was thought that the beams were one piece (110ft) and police escorts were being mentioned which resulted in an immediate panic attack; my solution was, 'You've got acetylene torches so put them to good use'. Luckily we did not need to and Nev Martin was on hand and knew just the right company to handle the transport. His help was invaluable as always. The Walton family who own Bruntingthorpe Airfield and latterly, of course, the Vulcan and a Victor tanker, have a common objective and love of aircraft preservation and have supported the LPG from day one. In the longer term a museum is planned and hopefully the QRA Shed will be an important part of this venture.

As a family the Waltons possess a variety of skills, which the LPG all too frequently have to call on as a favour. John Walton has a great deal more than my very limited knowledge of buildings and did not hesitate to join Jim Featherby and myself on our recce to Wattisham. We were greeted by Peter Booth and Peter Long of Trafalgar House and the sub-contractor, Brian Gale of Gale Construction, was also present as his company would be carrying out the actual dismantling. Work would start almost immediately with the removal of some of the cladding during the week commencing 20th December. A small team from the LPG would visit Wattisham during the Christmas break to identify and mark the main steelwork so that the task of reconstruction would hopefully be made easier. The first two of the six trailers required arrived at Wattisham prior to Christmas which would allow loading to take place without double handling. The schedule drawn up allowed just two weeks of January for completion of the job. The weather was less than kind and Wattisham was hit with snow in early January, which inevitably delayed progress, but an improvement in conditions resulted in our trailers removing the building between the 13th and 28th January. And so to the next problem. How do you arrange to unload six trailer-loads of steel? Fortunately the Waltons own some heavy metal in the form of a wide variety of "plant". Again we had to beg a favour and were indebted for loan of a very heavy duty tractor with a substantial fork attachment. Progress was sometimes painfully slow as much of the steelwork was also of the heavy duty variety, but the building is now in kit form at Bruntingthorpe. Fundraising has now begun in order to finance the application for planning permission and for reconstruction, and your purchase of this book will help us achieve our aim. It will be quite a challenge, but the thought of two F.6's in a Q-Shed for all to savour and enjoy will drive us on. It will be a very fitting memorial to a fine aircraft.

Dismantling of the Wattisham QRA Shed in January1994 [Bob Tuck]

LIGHTNING DEVELOPMENT

Over the years, a small group of pilots was largely responsible for Lightning experimental, developmental and test flying. They included Roland Beamont, Jimmy Dell, Desmond de Villiers, Peter Hillwood, Johnny Squier, Don Knight, Tim Ferguson and John Cockburn.

From the beginning, and following on from the successful Canberra flight test programme, English Electric Chief Test Pilot Wg Cdr "Bee" Beamont was responsible for the majority of the prototype and variant first flights and flight envelope expansion testing, and his first-hand accounts make fascinating reading. (Many are listed in the Bibliography.) "Bee" paid the Lightning Preservation Group the honour of taxying XR728 'JS' in June 1990 and very kindly agreed to write the Foreword for this book.

Wg Cdr Jimmy Dell was originally the RAF Project Pilot on the Lightning and was seconded to EE at Warton to assess how well the developing weapons system would fit service requirements. In late 1959 he was invited by "Bee" to join EE as Deputy Chief Test Pilot to take over from "Dizzy" de Villiers, who was unfortunately in poor health. Jimmy later became Chief Test Pilot upon "Bee" being appointed Manager of Flight Operations/Head of Test Pilot Flying. In June 1995 Jimmy also honoured the LPG by taxying XS904 at Bruntingthorpe, and kindly agreed to be interviewed for this book. Several of his accounts are included in this résumé of Lightning development to provide a flavour of the trials and tribulations of Lightning Test Flying.

All who love the Lightning owe a debt to those at Preston, Filton, Hucknall and elsewhere who played their own individual parts in the designing, building, testing, maintenance and operation of our favourite fighter.

LIGHTNING MARKS

P.1A

First flight 4.8.54
WG760, WG763 (and WG765, static test airframe)

P.1 WG760 is readied for flight at Boscombe Down, July 1954 [BAe]

The English Electric P.1 transonic research aircraft first flew on 4th August 1954 from A&AEE Boscombe Down. The aircraft was initially designed in response to Experimental Requirement ER.103 of 1947 by Teddy Petter and the Warton team, and the design was completed to Specification F23/49 which required a potential fighter configuration. Uniquely, its two 7,200lb thrust Armstrong Siddeley Sapphire unreheated turbojets were vertically stacked behind the pilot and staggered to minimise the aircraft's frontal area. Other innovative features were the 60° sweep 'notched delta' wing planform, with ailerons at 90° to the airflow, and the very low set tailplane; both of these features were regarded by RAE Farnborough as potentially dangerous and the Short SB.5 test aircraft, which was essentially a scale model of the P.1, had to be test flown with varying angles of wing sweep and different tailplane positions before they were satisfied that indeed English Electric had got it right. The Warton team were also pioneers in the use of leading edge slots instead of wing fences to prevent leading edge vortices breaking away outboard and affecting aileron effectiveness. The high thrust to weight ratio and low transonic drag enabled Chief Test Pilot Roland 'Bee' Beamont to achieve Mach 1.0 in level flight with ease during the third flight of WG760 on 11th August 1954, the P.1 being the first British aircraft with this capability.

First flight of P.1 WG760: 'Bee' passes Boscombe tower on 4th August 1954 using both leading and trailing-edge flaps. The leading-edge flaps were later locked shut as they gave no significant increase in lift.

'Bee' makes a close approach to the photographic Meteor on an early flight from Boscombe Down, demonstrating his confidence in the P.1's precise handling capabilities.

'Bee' banks away from the chase Meteor and enables a comparison to be made of the original wing plan-form (left) and the enlarged wings with conical camber (right) which were fitted to WG760 in 1956. [All photographs British Aerospace]

WG763 being fuelled prior to ground running at Warton. The new 250 gallon ventral fuel tank is clearly visible. [BAe via Jimmy Dell]

Pending availability of the Rolls Royce Avon engine, in 1956 a crude fixed-nozzle reheat system was fitted to increase thrust on each Sapphire to 9,200lb and so enabled the P.1 (now designated P.1A) to attain Mach 1.53 (approx. 1000mph). However, the fixed wide-open nozzles reduced 'dry' engine thrust to only 5,500lb, making single-engined landings problematic due to lack of power for an overshoot. The P.1A had only 574 gallons of usable fuel and this limited the speeds which could be reached before the aircraft had to return to base.

A rare photograph of static test airframe WG765 undergoing vibration tests. *[British Aerospace]*

WG760 at Binbrook in 1987. Centre panel: ASI top left, altimeter beneath; Machmeter missing on left hand panel; right hand panel contains engine instruments. *[Crown Copyright/MOD]*

In 1957 WG760 test-flew an enlarged wing with cambered leading edges and extended tips in preparation for tip-mounted stores, and this necessitated a small reduction in aileron span. These changes improved handling for the proposed all-weather role by raising the buffet threshold, improving the ability to pull high 'G' confidently and with precision, giving greater pitch control in the landing configuration and enabling the final turn to be made without buffet. In addition, drag was decreased subsonically (increasing estimated ferry range by 20%) although supersonic drag increased slightly. Unfortunately these wing modifications were not incorporated on production aircraft until 1965 with the advent of the F.6. As is well known, Lightning development was continually underfunded as the infamous 1957 Sandys Defence White Paper envisaged it to be the last RAF manned fighter before the Missile Age.

Jimmy Dell recalls the small cockpit of the P.1: "On my first take-off, there was suddenly a terrific vibration. At first I didn't know what the hell was happening, until I realised that the plastic hand-hold on top of my visor was touching the canopy. It then became clear why 'Bee' flew without his bone dome!" 'Bee' took second prototype WG763 for her maiden flight on 18th July 1955 and in September flew her as the P.1 debutante at Farnborough. WG763 was employed on performance and handling trials with a 250 gallon (2000lb) ventral fuel tank, and with two 30mm ADEN cannon mounted in the upper nose it carried out the first supersonic gun firing trials in this country. American-style toe-operated brakes were also tested but the traditional handbrake and rudder bar system was ultimately deemed preferable, both by English Electric and by Boscombe Down.

WG760 is now preserved, with its original wing planform, at RAF Cosford Aerospace Museum while WG763 enjoys well-deserved retirement in Manchester's Museum of Science & Industry.

P.1B

First flight 4.4.57
XA847, XA853, XA856

The P.1B was the aerodynamic prototype of the Lightning and was quite a different aircraft to the P.1A. F23/49 Issue 2 (3rd June 1953) specified a maximum speed of 1.7M and a rate of climb such that the time from pressing the first engine start button to reaching 50,000ft was to be 6 minutes or less. The cockpit was raised to improve rearward vision, airbrakes were redesigned and repositioned from between the jet-pipes to the upper rear fuselage and the simple 'pitot' air intake of the P.1A (which English Electric employed due to problems that contemporary fighters were experiencing with wing-root intakes) was redesigned with a central fixed 'bullet' to be suitable for higher speeds and to provide space for the Ferranti Airborne Intercept AI.23 radar. (The technology of AI.23 was also used in the Blue Parrot radar fitted to Buccaneers.)

The more powerful Rolls Royce Avon RA.24R (200 series) engines were now available for fitting to the P.1B. Each provided a dry thrust of 11,250lb, or 14,430lb with the 4-stage reheat system, and gave these lightweight (less than 26,000lb) prototypes a previously unheard of power-to-weight ratio of 1 to 1 at low fuel states. 'Bee' was very impressed with the Avon engine installation. Apart from the increase in thrust, there was a considerable reduction in noise and vibration levels and engine speeds could be widely desynchronised without a 'beat' being apparent. Some problems were being encountered with the lighting and combustion stability of the reheat system however.

'Bee' Beamont pilots the first P.1B XA847 on its maiden flight, during which Mach 1.13 was achieved. Note the small 'Stage 1' fin. [British Aerospace]

The P.1B became the Lightning on 23rd October 1958 when the first prototype (XA847) received a champagne christening by Chief of the Air Staff Sir Dermot Boyle. In the hands of 'Bee' Beamont, with the ventral tank providing the necessary fuel reserve, XA847 became the first British aircraft to exceed Mach 2 in level flight on 25th November 1958. This eclipsed the British air speed record of 1,132mph held by Peter Twiss in the Fairey Delta 2 (also designed in response to ER.103), but the P.1B made no official record-breaking attempt as it still held insufficient fuel to perform the required high-speed run in each direction! It was planned that operational Lightnings would have internal fuel capacity increased from 574 to 710 gallons (4500 to 5500lb) by using almost all available space, including converting the split flaps into single units and filling them with fuel.

A DH Firestreak missile ready for launching from a P.1B fuselage rig at RAF Valley in Anglesey. [British Aerospace]

At this time the Lightning was seen as a point-defence interceptor whose primary role was to defend V-bomber bases against a high-speed, high-altitude bomber threat. The early marks of Lightning, F.1 and 2, were to use currently available weapons and systems pending the availability of the F.3 which was to be the ultimate interceptor. The 1957 White Paper dictated that the Lightning's primary operational role was to lie between 40,000 and 60,000 ft, i.e. above SAM missile range. This dictated the Lightning's destiny: its performance in this role was not to be degraded by performance considerations elsewhere in its operating envelope. This may be one reason for the delay in introduction of the cambered wing, which slightly increased supersonic drag, until 1965 when low-altitude operations and long-range ferry flights were beginning to increase in importance.

XA847 fitted with the larger 'Stage 2' fin and an early mock-up of the enlarged ventral tank. Black painted nose and the fuselage mounted camera are evidence of flight refuelling trials. [BAe]

Trials with two Blue Jay (DH Firestreak) missiles and with the ventral tank demonstrated a significant reduction in longitudinal stability and it became obvious that a fin of larger area was required, together with the introduction of autostabilisation. XA847 was used for trials of the 30% larger Stage Two fin (later fitted with a forward extension on this prototype only), and also trialled mock-ups of proposed alternative missiles including the US Genie nuclear-tipped unguided air-to-air missile. A maximum of 240 Genie missiles were to be supplied free under US aid but the requirement was cancelled in January 1960. Falcon and Sparrow guided missiles were also considered but were incompatible with AI.23.

Encouraged by the Canberra's export success, Australia was offered a variant of the proposed Lightning F.3 in 1960, although it was suggested by English Electric that the two-man crew of a T.5 would be better able to exploit the capabilities of the upgraded weapon system. An enlarged ventral pack was proposed containing extra fuel and doppler navigation and with the missile pylons towards the rear. The aerodynamic shape was trialled on XA847 & XN725 and was more cylindrical than the later area-ruled F.6 ventral fuel tank. Up to 4 Sidewinders were also proposed for interception but unfortunately this study did not come to fruition and a major opportunity to upgrade the Lightning's capabilities was missed, to the profound regret of successive generations of Lightning pilots. For ground attack, rockets, HE bombs, nuclear weapons or US Bullpup pilot-controlled missiles were also proposed, but unfortunately the Australian export venture came to nought.

Following runway over-run taxi trials at RAE Farnborough and at the end of its useful life XA847, a truly historic British aircraft, was displayed at the RAF Museum, Hendon until sold to Wensley Haydon-Baillie in 1988 to make room for XS925, a Mk. 6 which would be more representative of RAF Lightnings. XA847 was reassembled at Farnborough for the 1994 SBAC Show, but is reported now to be back in store at Southampton. Its long-term future is not known.

A TASTE OF THE P.1B

Squadron Leader PETER CARR AFC RAF (Retd.)

I was briefly involved with the P.1B during a tour of duty with the Air Fighting Development Squadron (AFDS) at West Raynham in 1958, when the 'powers-that-be' decided it would be a good idea for three current fighter pilots to make a tactical evaluation of the P.1B before it went into production in its final form as the Lightning.

Wg Cdr Colin Coulthard (O/C AFDS), Wg Cdr Jimmy James (O/C All-Weather Development Squadron, also at West Raynham) and myself were selected and spent a considerable amount of time at E.E. Warton with Roly Beamont, Jimmy Dell and the many others involved with the Test Programme. However, actually getting to fly the beast became a major problem. The weather factor at that time of year, together with the somewhat limited landing aids, curtailed the flying activities for us 'new boys' in particular. These limitations, together with a very busy test flight programme requirement, resulted in my achieving only three flights each of approximately 30 minutes duration over the period 17th Jan—2nd March 1958. I believe Colin Coulthard also managed three flights and Jimmy James a few more.

Obviously this limited exposure over such a protracted period did not allow more than a general impression of the P.1B, but my lasting memory is of its phenomenal acceleration and rate of climb, together with its general good handling, which far exceeded my experience of the F.100 whilst on exchange with the USAF in America.

I nearly had a calamity on my second flight when trying to land in very restricted visibility. It was very hazy on take-off and whilst undertaking a supersonic exercise over the sea Warton recalled me urgently as a sea fog was rapidly covering the airfield. Fuel reserve was very limited as total endurance after lengthy use of reheat was some 30-35 minutes in that configuration. Unfortunately there was no sign of the runway on my first approach and there was only enough fuel for one overshoot.

The visibility was now extremely poor and I was not completely happy operating at low level in an unfamiliar aircraft, with no knowledge of local terrain and the imminent possibility of a flame-out. On the second (last) approach with maximum help from the Air Traffic Controllers, lots of flares and a lot of luck, the airfield boundary appeared at the same time as I saw the runway some 100 yards to the right. Fortunately, there was sufficient space for me to correct and touch down far along the runway and thankfully bring the aircraft to a halt helped by an efficient drag 'chute plus red-hot brakes. The thought of losing one of the limited number of prototype aircraft still makes me sweat.

In 1959 I was fortunate enough to be given command of 74 Sqn at Coltishall with Hunter Mk.6 aircraft. Soon thereafter I was told that we would be the first squadron to be equipped with Lightnings, but by then I had asked to be released from the RAF as Donald Campbell had asked me to join him as Project Director and reserve driver on his Bluebird projects. I therefore left the RAF early in 1960 with much lingering regret and jealousy every time Lightnings were in the news.

Roland Beamont banks XA847 over the Ribble estuary in the circuit at Warton having just selected gear down. [British Aerospace]

A CLOSE CALL

JIMMY DELL OBE

Formerly RAF Lightning Project Pilot and later English Electric Chief Test Pilot

The closest escape I had (even counting my ejections from XM966 and an F.86E) was in XA853 on 8th May 1958. I took-off to do some gunnery attacks on a Hunter. I was supposed to meet him at 40,000ft over the Irish Sea and was climbing up through cloud when there was a loud BANG. I didn't know what had happened but the aircraft began flying slightly sideways. (Later on I found out that a defective jet pipe had burst and blown open the left airbrake; it had also flattened the air-duct at the base of the fin, so there was no cooling air to the electrical generation system and hot gases were also pouring on to it.) Without electrics I had no indications in the cockpit, no lights, no radio. Obviously something fairly serious had occurred and so I immediately returned to base.

[BAe]

I tried to fly down the runway but missed slightly because I was going sideways. At that time we didn't have any rear-view mirrors so I couldn't see if there was any smoke. I didn't have fire warnings so I crabbed on to the downwind leg and quickly selected undercarriage down on the emergency system. When the wheels go down there is a good solid *THUMP, THUMP* of the main gears and after a pause a *THUMP* which is the nose-gear. After a short wait I heard: *THUMP, THUMP*—pause—*THUMP*. Thank God! And then it went *THUMP* a fourth time—sounding just like the nosewheel—so that threw me into a bit of confusion! However, there was no way I could go round again without knowing what damage had ensued.

As the wheels touched I streamed the brake 'chute and as I came to a halt the controls seized completely. I had no hydraulics any more as all the fluid had gone out of the back, and the aircraft was on fire although I didn't know it at the time. The fire trucks came down the runway towards me and stopped: I tried to attract a fireman's attention but he disappeared down towards the back end, and eventually I had to shout for a ladder as they were more concerned with extinguishing the fire than getting me out! If the controls had seized about 20 seconds earlier whilst on the approach then I would have lost control, and been in the same situation as George Aird who ejected from XG332 at Hatfield.

Sister-ship XA856 at Rolls Royce Hucknall in April 1958 with original fin and two high-spirited pilots in an Austin 7.
[Rolls Royce plc]

Fortunately, XA853 survived and was rebuilt to carry on development flying.

F.Mk.1 Development Batch

First flight 3.4.58
XG307-XG313, XG325-XG337

To speed up clearance to operational standard, 20 pre-production Lightnings were built to work up individual systems in parallel, such as radar, guns, missiles, radio, TACAN, and to undertake handling, tropical trials and so on; after the first three DB aircraft the larger Stage Two fin was fitted. In parallel with the Ferranti AI.23 radar, sponsored by RAE, the Royal Radar Establishment designed the AI.20 to be built by Ekco. Seven development batch aircraft were to be fitted with AI.20 and five with AI.23, but unsatisfactory performance of AI.20 led to its cancellation with resultant delays and extra design time while the Lightning programme was rearranged to accommodate AI.23. The majority of the AI.23 radar trials were flown in XG312 and XG327.

XG336 with 'Stage 2' fin and four gun ports leaving Warton's No. 2 hangar. Behind are 7 single-seat aircraft and a T.4 all apparently undergoing modification. [British Aerospace/North West Heritage Group]

It was not possible to use TACAN as an effective navigation aid throughout much of the Lightning development programme and so all-weather navigation was achieved with the assistance of Warton and Ulster Radars. The TACAN equipment was designed in the USA and the UK Ministry of Supply spent £5M building it in Britain. Unfortunately this first 'A' version was hopelessly unreliable, the improved 'B' version was almost as bad and eventually MoS was forced to buy a completely redesigned 'C' version from America.

In December 1959 three aircraft were delivered to the AFDS, now at Coltishall, and on 16th June 1962 "Bee" took XG310 on its first flight after the fitting of the even larger Stage Three (145%) 'square-top' fin which was to become standard on the later marks. Surviving Lightnings from the Development Batch are XG329 at Flixton, Suffolk, which trialled the guns and Datalink, and XG337 at RAF Cosford whose trials included Auto ILS and Red Top missiles.

DB Lightning arriving by sea at Aden for hot weather trials, 1961. [Crown Copyright/MOD]

(above) The wreckage of XG311 is recovered by a Belvedere from the sea off Lytham.
Don Knight had ejected successfully after an undercarriage malfunction.
[Lancashire Evening Post]
(below) P.1127 XP980 is flanked by three DB Lightnings at A&AEE Boscombe Down in August 1967.
To the left are XG336 and XG337 with the final 'Stage 3' fin design.
[Crown Copyright/MOD]

F.Mk.1

First flight 29.10.59
XM134-XM147, XM163-XM167 (& XM168, static test)

On 29th October 1959 Roland Beamont took to the air in XM134, the first of the production batch of the Lightning F.1 and similarly powered by RR Avon 200 series engines with four-stage reheat. This was the operational version of the P.1B and was equipped with two 30mm ADEN cannon housed in the large 'elephant-ear' equipment bays behind the cockpit and firing through troughs in the upper nose behind the intake ring.

The Ferranti AIRPASS (Airborne Intercept Radar Pilot Attack Sight System) AI.23 radar was housed in the pressurised nose bullet and used to control DH Firestreak infra-red pursuit-course guided weapons mounted on an exchangeable weapons pack beneath the forward fuselage. In its final version, the AI.23D radar was designed to search for the target indicated by ground control then, when the target was acquired, the fire control system computed steering instructions (according to the type of weapons selected) which appeared on the radar attack display as a steering dot to be maintained within a reference circle. Slaving signals were also supplied to the missiles to enable their homing systems to acquire the target before launching, and this was indicated by two of the four event markers at the top of the radar scope. Computers would calculate firing range brackets appropriate to the selected weapons and these would be displayed as a reduction in the diameter of the time circle (a larger circle concentric with the reference circle). The first reduction of diameter occurred when the target was within range and the second when the missile automatically fired. The breakaway signal was a large cross on the screen which was the cue for immediate evasive action to avoid debris. As an alternative to guided missiles, English Electric made provision for 48 (later reduced to 44) x 2" high-explosive unguided rockets in two retractable Microcell packs (although these were not used by the RAF) and 'in-range' firing signals for these too were indicated on the AI.23 display.

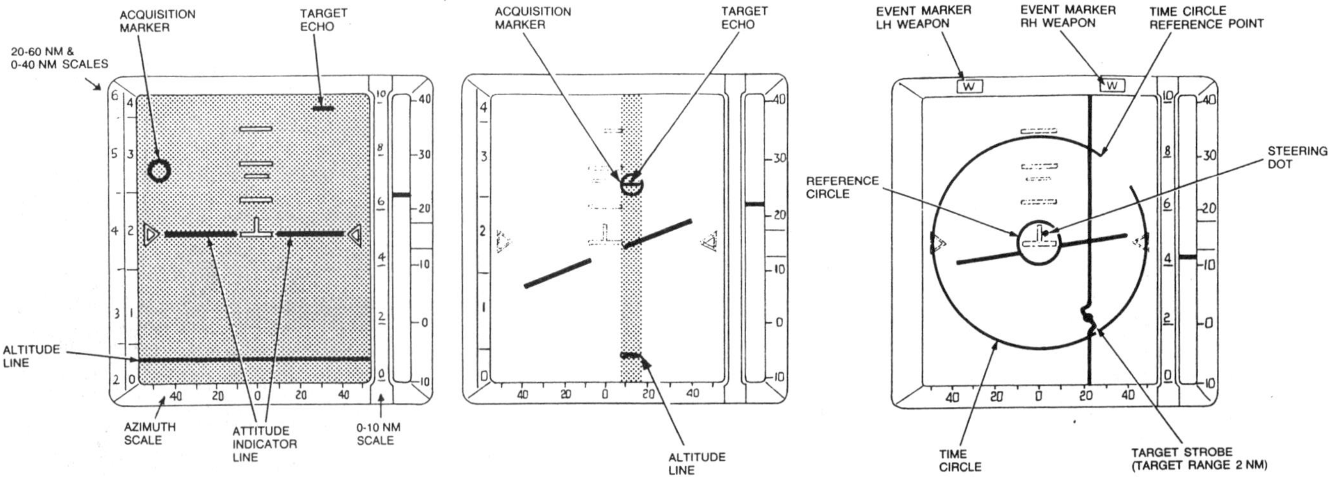

Search Phase | Acquisition Phase | Track Phase (10 n.m. scale)

<u>AI.23D Cathode Ray Tube (CRT)</u>: The screen is flanked by 3 vertical scales, 20-60 nm and 0-40 nm to the left and 0-10 nm range to the right. At the bottom is a horizontal azimuth scale, up to 50° port/stbd. The acquisition marker is a small circle which can be positioned in azimuth and range by the AI Hand Controller on the port console; it is placed over the target echo in order to determine range information which can be used by the radar and weapons 'red computer'. During the Track phase the radar locks on to the target and the target echo appears in the centre of the tracking strobe, an S-shaped kink indicating the range. The steering dot and small central reference circle appear. When the dot is tracked by the circle the computed attack course is flown. The time circle is a larger circle concentric with the reference circle, a gap in the circle giving an indication of speed relative to the target: if the target is being overtaken the gap in the time circle rotates clockwise from '12 o'clock' (each 90° of arc representing 1000 knots) or anticlockwise to give the reverse or opening speed. Two weapon event marker panels display a W as the related port or starboard weapons acquire the target. When missiles are to be launched the diameter of the time circle reduces to 2/3 ("in-range") and then 1/3 of its diameter (indicating "last chance to fire" for Firestreak or "automatic firing" for Red Top missiles). [British Aerospace]

One particular Firestreak trial is recalled by Jimmy Dell: "I did one of the early Firestreak firings when I was fairly new to the job at Warton. It was at Boulmer range out over the North Sea and I had to accelerate to 1.4M, 36,000ft and then zoom climb to 0.94M at 50,000ft. The Test Point was to fire a single missile off the port side while in a 2G starboard turn at 0.9M, 50,000ft. I was briefed by a young flight test engineer and I thought, well, I've never pulled 2G at 50,000ft but if he says so then the aircraft must be capable of doing it.

"You have what is called a Safety Trace, whereby they say: 'Right, you start the turn *here* and you've got to loose off the missile in that arc of turn; if you get to *here* and it hasn't gone, then don't fire as it will go outside the range safety area.' This didn't leave me much time to pull on the 'G' and as I started to pull round to starboard the drag of the single missile on the port side was causing me to sideslip, so I had to apply quite a lot of rudder to hold it. (I learned later that you should anticipate the situation, apply rudder before sideslip occurs and then ease off so that you are fully in control.)

"I had only got to about 1.6 or 1.7G but the side-slip was increasing; so I fired the missile, whereupon the aircraft suddenly departed to port. It rolled, the nose dropped and the next minute I was looking at the sea! Realising it had stalled, I centralised the controls and eased out. Normally the missile would have gone out of camera-shot fairly quickly but because the aircraft flipped left as I fired it remained in the camera's field of view—it looked as though I had deliberately rolled around the firing axis, when in fact I was out of control! I was quite excited when I got back and Freddie Page, who was then Chief Executive, asked what the firing conditions had been. When I said '2G, 0.9M at 50,000ft' he shook his head: 'No, no, not possible!'. So that was an early lesson I learnt, to double check what I was told."

Lightning major components were built at the English Electric works at Preston Strand Road (front fuselages and wings) and at Accrington (rear fuselages and tailplanes), with plastic components such as canopies and nose cones made at Freckleton adjacent to the hangars at Warton. Final assembly took place at Samlesbury and upon completion of ground testing the first flight was the delivery to Warton on the other side of Preston for flight testing. Maximum production rate of the F.1 was 3 aircraft per month, but this increased to 4 aircraft per month for the F.1A.

(left) Skinning stage for the front fuselage of F1A XM184.
(right) Front fuselage of F1A XM177 in the turnover rig.
[Both British Aerospace]

XM134 was delivered to Boscombe Down on 31st March 1960 for acceptance testing, whilst XM135 was delivered to the Central Fighter Establishment at Coltishall on 14th May, followed by XM136-8 over the next few weeks. Also based at Coltishall was No. 74(F) 'Tiger' Sqn, by then commanded by South African Sqn Ldr John Howe, who relinquished their Hunters and began to work up on the Lightning despite having no two-seat trainers. First delivery to 74 Sqn took place with XM165 on 29th June 1960 and four aircraft were demonstrated at Farnborough in September. The Squadron was up to strength by January 1961, declared operational shortly afterwards, and went on to display a six-ship formation at Paris in June and a nine-ship at Farnborough in September 1961. In 1962 The Tigers celebrated their role as Fighter Command aerobatic team by painting the fins and spines of their Lightnings gloss black, foreshadowing the exotic Lightning colour schemes which appeared during the 1960's.

The original Service limit of Mach 1.7, imposed in 1960 pending completion of flight clearance of the air intake, led to rumours at the 1961 Paris Air Show that intake buzz limited the Lightning to this speed. 'Bee' countered the propaganda by stating that throttle slams from maximum reheat to flight idling had regularly been demonstrated at speeds in excess of M2.0 without intake buzz and affirmed that, with good design, variable intake geometry was not necessary for speeds up to Mach 2.0.

The only complete surviving F.1 airframe is XM135 at IWM Duxford, well-known as the vehicle for Taff Holden's unintentional flight at Lyneham which has been extensively documented elsewhere (such as in John Beattie's book 'Drama in the Air').

(above) Thumbs up from Sqn Ldr John Howe in XM143 'A' at a photo-call to celebrate entry into service with 74 'Tiger' Squadron. [British Aerospace]
(right) Lightning F.1's of 74 Squadron in echelon starboard [Rolls Royce plc]

F.Mk.1A

First flight 16.8.60
XM169-XM192, XM213-XM216

The F.1A was an improved version of the F.1 with UHF radio in addition to VHF, improved windscreen rain dispersal and provision for in-flight refuelling; it could be distinguished externally from the F.1 by the presence of cable-ducting on the lower fuselage. Deliveries to 56(F) Sqn began with XM172 on 14th December 1960 and in March 1961 deliveries began to 111(F) Sqn, both squadrons being based at Wattisham. No. 56 Sqn became the official RAF aerobatic team 'The Firebirds' in 1963 *(see Jim Jewell's article)* and performed at the Paris Air Show. At least four F.1A airframes (XM172, XM173, XM178 & XM192) are known to survive of the 28 built, although XM173 was rumoured to be up for tender at the time of writing.

F.Mk.2

First flight 11.7.61
XN723-XN735, XN767-XN797; some later converted to F.2A or F.52

The F.2 had fully variable reheat for its RR Avon 210's and a liquid oxygen (LOX) system. It also introduced Phase 1 of the OR.946 Integrated Flight Control and Instrument System, which processed data from the Master Reference Gyro and Air Data Computer for display on the primary servo-operated flight instruments. The navigation display could be switched between Compass mode, ILS (or UHF homer), TACAN and Datalink (DL), although the latter never entered production. The Flight Control System was based on the Flight Control Computer. Flight Director presentation of a course to steer manually was given on the radar 'B'-scope and could incorporate programmed climb, height and heading lock, attitude hold and ILS track and glide. Alternatively selection of Autopilot enabled the desired course to be flown automatically. The planned fit of OR.946 into the F.1 had previously been abandoned due to delays in equipment delivery which were largely attributable to its development in a piece-meal fashion rather than as an overall system. Similar holdups were largely responsible for the long delay (over 18 months) between first flight and delivery into service of the F.2. To minimise delays, aircraft were completed to a sub-standard at Samlesbury and flown to Warton to be stripped down and rebuilt to final modification standard.

Before and after: The first F.2 XN723 on the production line at Samlesbury in 1962, and in March 1964 after crashing near Keyham due to an engine fire. Pilot D. Withnal ejected safely. [British Aerospace/Rolls Royce plc]

Externally the F.2 could be distinguished from the F.1A by the mid-spine intake for the stand-by generator, although more obvious features were the 'XN' serial and usually the presence of 19 or 92 Sqn markings, to which squadrons F.2's were first delivered in early 1963 prior to the squadrons moving from Leconfield to RAF Germany in 1965. Two additional cannon could be fitted in place of the missile pack, but these are not known to have been used operationally and the troughs under the nose were later faired over. No intact Lightning F.2's exist but fortunately the Saudi government has preserved two of the similar F.52 airframes at Dhahran and Riyadh.

F.2 XN773 in flight with a Spitfire in October 1964. AFDS Lightnings took part in combat trials with Spitfires in 1963 to develop tactics in case Lightnings had to engage Indonesian P.51 Mustangs - poor targets for Infra Red missiles. [Rolls Royce plc]

F.Mk.2A

XN724, XN725 [to F.3A], XN726-8, XN730-3, XN734 [F.2A/3A hybrid], XN735, XN771-8, XN780-4, XN786-793, XN795 [partial conversion]

F.2 XN734 was partially modified to F.3 standard and used by Rolls Royce for Avon 301 engine development. It is shown here after leaving the runway due to a brake 'chute failure in October 1966. In 1967 it was modified to take the large ventral tank of the F.2A but retained the straight wings. It was eventually used by BAC as a ground trainer for the Saudi contract, coded G27-239, and was finally scrapped at Cranfield on 1st December 1994. [Rolls Royce plc]

Most F.2's were reworked by BAC to F.2A standard, having F.6 features such as the enlarged (flat-top) fin, spring steel arrestor-hook and cambered wing leading edge for improved low-speed handling. F.2A's entered service on 15th January 1968 with 19 Sqn, and perhaps this variant should have been designated the F.7 as it was the final RAF variant and favoured over the F.6 by most pilots: the F.2A was lighter than the F.6 as it was not strengthened to take overwing tanks and it had two fixed cannon in the upper nose which gave the additional advantage that all of the enlarged ventral tank could be used for fuel. Although slightly lower-powered, its RR Avon 301's were smaller and easier to change than the 302's of the F.6 and it had a very reliable spark-plug reheat system, unlike the temperamental 'hot-streak' system of the F.6. Pursuit-course Firestreaks were the only missile option but they again tended to be more reliable than the later collision-course Red Tops. The only remaining F.2A airframe in the UK which can truly be described as preserved is XN776 at East Fortune, but three more exist in Germany.

F.Mk.3

Production aircraft first flight 16.6.62
XP693 [to F.3A/6], XP694-6, XP697 [to F.3A/6], XP698-708, XP735-765, XR711-721, XR722 [to F.53], XR723-728 [to F.6], XR747 [to F.6], XR748-751

The F.3 was initially fitted with RR Avon 301's which each gave a thrust of 12,690lb dry and 16,360lb with reheat, and improved AI.23B fire-control radar for the Firestreak or more advanced Red Top IR collision-course missiles. As the nose guns were deleted, the two missiles were the F.3's only armament—widely regarded as a mistake, because apart from leaving the aircraft with minimal armament it removed the option of firing warning shots at a straying WarPac aircraft.

Hawker Siddeley Dynamics Red Top missile trials on DB Lightning F.1 XG325 at Hatfield. [British Aerospace]

OR.946 Attitude Display (lower left) and three modes of Navigation Display which could be selected (clockwise from top left): Compass, Instrument Landing System and TACAN. [British Aerospace]

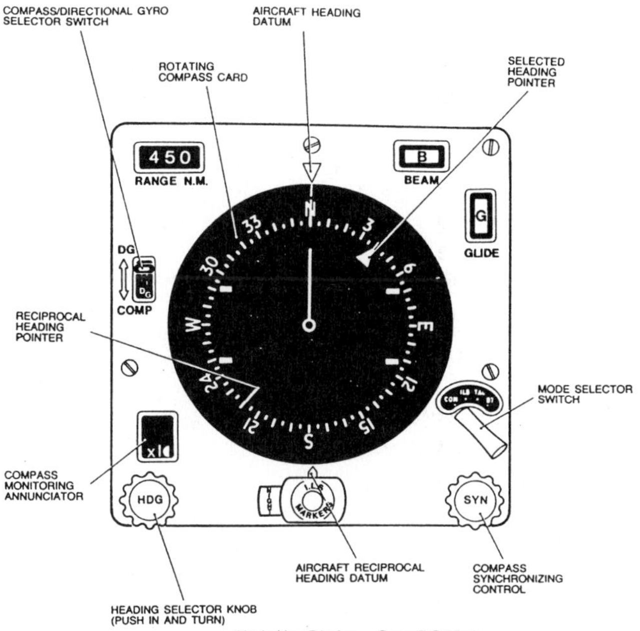

Navigation Display – Comp/DG Mode

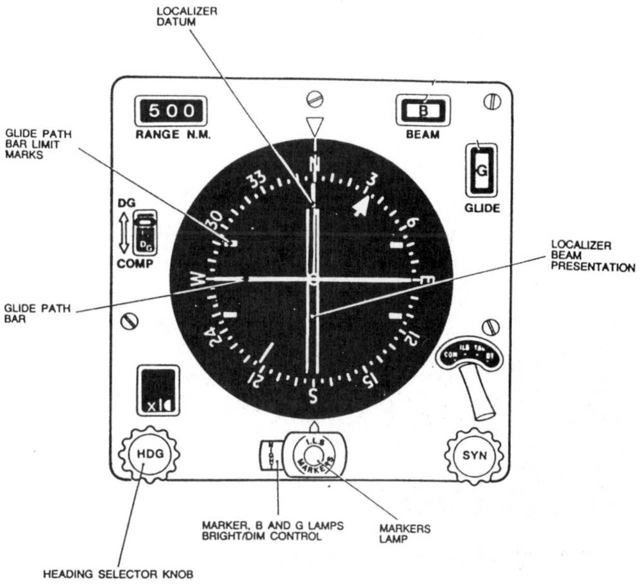

Navigation Display – ILS Mode

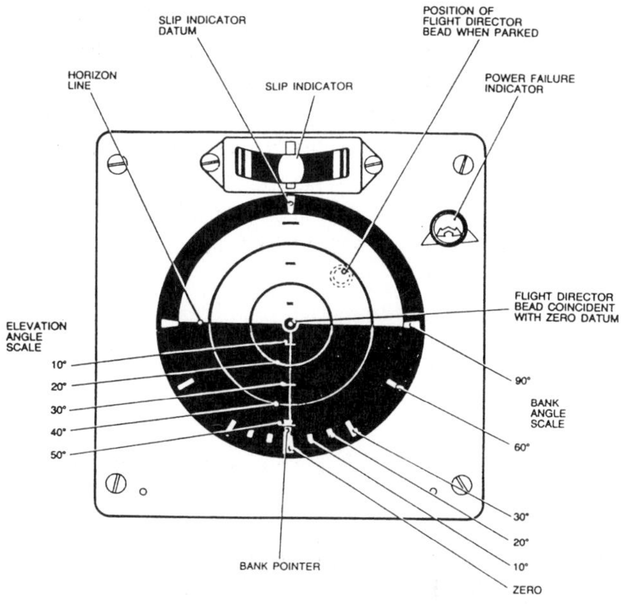

Attitude Indicator

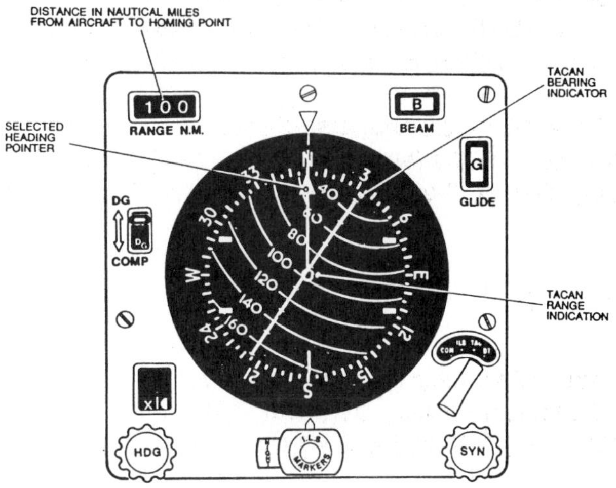

Navigation Display – TACAN Mode

Distinguishing features over the F.2 included forward extension of the fuselage cable ducting and the large flat-topped fin, introduced to improve longitudinal stability when carrying Red Tops at high speed. Phase 2 of the OR.946 flight control and instrument system was also introduced. First production F.3 was XP693 (Lightning enthusiasts' favourite of later decades!), retained by the manufacturers and later converted to F.6 standard. The F.3 first entered RAF service on 1st January 1964 when XP695 was delivered to the Central Fighter Establishment, and squadron service began on 14th April 1964 when XP700 joined 74 Sqn at Leuchars; however, delays in equipment delivery to English Electric became so acute that the Ministry decided to store 12 F.3's to ease the situation. Out of the 62 built which were not subsequently converted to F.6 or F.53, at most six complete F.3 airframes survive: XP706, XP745, XR713, XR718, XR749 & XR751.

F.3 XP755 'E' of 29 Squadron on APC at Akrotiri in 1974
[Crown Copyright/MOD]

T.Mk.4

First flight 6.5.59
XL628-9, XM966, XM968-974, XM987-997

The P.11 or Lightning T.4 was a fully-operational two-seat trainer version of the F.1A and managed to accommodate instructor and student side-by-side, even though the cockpit was only 11.5" wider than that of the single-seater and could be described as 'cosy'. Labour shortages led to T.4 front fuselages being built in the Experimental Shop rather than using normal production shop methods. On 29th June 1962, XM970 was the first T.4 to enter RAF service with the Lightning Conversion Squadron formed at Middleton St George, which in 1963 became 226 OCU. The Lightning T.4 was operated both in the UK and in RAF Germany and although it lacked guns it has been described as the most potent air combat aircraft of its day, having the advantage of a second pilot who could act as systems operator. (The later T.55, not purchased by the RAF, was even better due to its longer range and heavier armament.)

'Bee' approaches the Meteor chase in the first prototype T.4 XL628, which was later lost when the fin broke off. [BAe]

The first prototype T.4, XL628, was lost on 1st October 1959 when the fin failed in a violent yaw at Mach 1.65 during roll-coupling evaluation. Johnny Squier ejected whilst supersonic and then, due to a failure of his rescue beacon, spent 28 hours in a dinghy before dragging himself ashore on the Scottish coast. Second prototype XL629 was more fortunate, after ETPS use at Boscombe Down it was assigned the more sedate role of gate-guardian and is the only T.4 still surviving. Long may it remain.

(above) A T.4 fuselage apparently joined only temporarily as the wings, which are bolted together before assembly, have yet to be inserted at the transport joint aft of the cockpit. [British Aerospace]

(above) Mark Lambert of Flight International magazine looks rather apprehensive about what Jimmy Dell is going to subject him to in a T.4. [BAe via Jimmy Dell]

(left) T.4's and DB F.1's on the line at Warton. Nearest aircraft XM993 overshot the runway at Middleton St George and burnt out on 12th December 1962. [British Aerospace]

T.Mk.5

Production aircraft first flight 17.7.64
XM966-7, XS416-423, XS449-460, XV328-9

The T.5 was in turn a trainer version of the F.3 and again could be used as a 2-seat operational interceptor—this was initially to be achieved by detaching the right-hand control column. The airframe was derived from the T.4 but performance was improved by fitting Avon 301 (later 302) engines, increasing fin area and raising equipment fit to F.3 standard. In the T.4 the instructor in the right-hand seat had the throttle in the conventional position on his left; however, in the T.5 the throttle was moved to his right, so in the event of taking over control from his student he had to fly and land the 'Frightening' left-handedly, which it was suggested might be a problem. Indeed, Peter Orme tells of an occasion where a pilot on a low overshoot from finals in the right-hand seat of a T.5 closed the throttles with his right hand (instead of easing the stick back) and pushed the stick forward with his left hand (instead of pushing the throttles open). Peter, in the left-hand seat, quickly pushed his throttles into reheat and snatched the stick fully aft; the aircraft bounced heavily off the main wheels as the reheats lit, but fortunately he managed to retrieve the situation without damage to the aircraft.

Development work on the T.5 was based around the F.3 programme, but specific T.5 flight trials would be carried out by the first two T.4 production aircraft which would be converted to T.5 prototypes. XM967 was stopped in build and scheduled to fly in July 1961 after conversion to T.5 standard. It was decided that XM966 would be the second prototype as it had already flown as a T.4 (15th July 1960, incidentally even before the F.1A) and so would take longer to convert; it was scheduled to fly as a T.5 in February 1962. XM966 would be a cheaper aircraft to convert than XL629, the surviving T.4 prototype, as the latter was not to production standard and had also undergone repair work following a minor accident on 7th July 1961.

XM966, T.5 first prototype, before and after its crash on 12th July 1965 due to fin failure. Just visible in the right-hand photograph next to the roundel are the open doors of the rocket pack. [right: via Trevor Garrod & Chris Brydges]

As English Electric Aviation was heavily involved with Lightning F.3 and TSR.2 work, parent company BAC decided that Bristol Aircraft Ltd at Filton would design and construct the T.5 nose fuselage, convert the two prototypes and supervise their initial flight testing at Filton before delivery to Warton. EEA would design and build the remainder of the airframe and would be responsible for T.5 final assembly and production. The rear fuselage was planned to be interchangeable between Marks 3 and 5, subject to modification of the forward spine area.

To avoid unspecified 'major structural changes' the T.5 prototypes uniquely employed electrical engine starting, rather than the conventional volatile starter fuel 'Avpin' (Isopropyl Nitrate, IPN). Jimmy Dell recalls that electrical starting required the use of a large van and this option was not used on other Lightnings. The T.5's were also planned to have nosewheel steering (NWS). This was later cancelled, although a Phantom NWS system with gearing adjustments was trialled on DB aircraft. The trial was a success and saved a lot of brake use, but the handbrake and rudder system worked satisfactorily on the squadrons and it was decided not to spend a lot of money on retrofitting.

It was also proposed that T.5's would have the option of carrying an air-to-air Microcell rocket pack (44 x 2" SNEB) in place of the guided weapon (Firestreak/Red Top) pack and have the facility to carry jettisonable overwing tanks or a pair of 68mm rocket-pods (2 x 18). This never happened on production T.5's but the rocket packs were offered for export Lightnings.

XM967 first flew on 29th March 1962, 8 months late, and XM966 had its first flight after modification on 1st December. Unfortunately '966 was lost on 12th July 1965 due to failure of the fin while engaged in rolling manoeuvres with rocket pack extended at Mach 1.8, 3G at 35,000ft, test pilot Jimmy Dell and test observer Graham Elkington ejecting safely into the Irish Sea. This repetition of the T.4 accident led to the larger finned Lightnings being further strengthened. By way of a postscript, a quarter of a century later on 28th April 1988 a small ceremony was held at Warton at which Jimmy Dell was reunited with the helmet he had worn upon ejection, which had been captured in a trawler's nets.

The first production aircraft (XS416, now preserved at New Waltham) was originally scheduled to fly by October 1962. It was estimated that twenty months would be required to build the aircraft (February 1961 to September 1962 inclusive) and Service delivery was planned for March 1963; however, delays resulted in XS416 first flying on 20th August 1964, and it was retained for development work by BAC and A&AEE until its delivery to 226 OCU on 26th July 1965. In the event, second aircraft XS417 (now preserved at Newark) beat XS416 into the air and on 17th July 1964 was the first production T.5 to fly. It was delivered to 226 OCU on 25th May 1965, although T.5 XS419 had already joined the OCU on 20th April. XV328-329 were attrition replacements built in 1966, almost a year after T.5 production had ended. Today, nine T.5's still survive in varying degrees of preservation.

Jimmy Dell stands in front of XS904 at Warton on 28th April 1988 after being re-united with the 'bone-dome' he wore when he ejected from XM966 23 years earlier.
[British Aerospace/North West Heritage Group]

F.Mk.6

First flight of XP697 as F.6 17.4.64
XP693, XP697, XR723-728, XR747 (converted from F.3). XR752-773, XS893-904, XS918-938

The F.6 was powered by RR Avon 302's and had the extended 'conical camber' wing first tested on WG760 in 1957. The Lightning's short endurance was addressed by the provision of overwing 260 gallon ferry tanks ('overburgers') and an enlarged area-ruled ventral fuel tank which actually reduced drag. It held 610 gallons, reduced to 535 gallons if the front part of the tank were replaced by a pack of 2 ADEN 30mm cannon, the guns being cooled by fuel. Maximum fuel load was 1,846 gallons (14,400lb). Although heavier than earlier marks, the F.6 could take off, climb to 36,000ft and accelerate to supersonic speed in less than 3 minutes.

The first prototype F.6 was XP697, converted from F.3 standard. Other new-build F.3's were delivered from Samlesbury to Warton for storage prior to F.6 conversion (this batch included XR728). First production Interim F.6 (lacking overwing tank capability) was XR752, this variant being known initially as the F.3A or F.3*.

Classic shot of XR754 over the Lake District. At this time the F.3A/F.6 was not equipped with an arrestor hook.
[British Aerospace]

The Interim F.6 entered service with 5 Sqn at Binbrook when XR755 and XR756 were delivered on 10th December 1965, followed by delivery of the first full-production F.6 XR768 to 74 Sqn on 1st August 1966, prior to the Tigers' deployment to Tengah, Singapore in June 1967. Despite the later serial numbers of T.5's XV328 and XV329, the honour of being the last Lightning built for the RAF goes to F.6 XS938, first flown 30th June 1967 and delivered to 23 Sqn on 28th August 1967. However several Lightnings retained at Warton for F.6 modification were delivered after this date, including XR728 to 23 Sqn on 1st November 1967. Surprisingly, the last new Lightning to reach the RAF was XR751, an unmodified F.3 which was delivered to 60 MU on 19th January 1968 and which still survives in Cornwall.

(above) Lightning F.3A to F.6 modification line in No. 4 hangar at Warton on 24th January 1968. XR764 is nearest the camera: it crashed off Cyprus on 30th July 1971 after a jet pipe fire, the pilot ejecting safely. [British Aerospace/North West Heritage Group]

(right) F.6 cockpit at Binbrook in November 1987 with seat removed; OR.946 instruments are in the centre. [Pete O'Callaghan]

The Lightning, whose design originated from a requirement issued in 1947, was finally stood down from RAF service with XI Sqn at RAF Binbrook on 30th April 1988. Even at this late date its high speed performance at altitude was still outstanding, but its meagre armament, obsolescent radar and worn-out airframe eventually led to its withdrawal after 28 years in the front line of UK Air Defence. Several airworthy airframes were put up for tender and the Lightning Preservation Group (LPG) was fortunate to have Station Commander Group Captain John Spencer's 'personal' mount XR728 'JS' delivered to Bruntingthorpe on 24th June 1988 by Flt Lt Chris Berners-Price.

Five Lightning airframes continued to be used by BAe Warton as high-speed targets for ADV Tornado AI.24 Foxhunter Radar Trials. After withdrawal of XR724 it was stored at RAF Shawbury before being acquired by the Lightning Association and ferried to Binbrook. The Trials Programme was completed by December 1992, following which the Lightning Flying Club/Classic Jet Aircraft Company acquired XP693 and XR773 for potential airshow use, stripped non-airworthy XS928 for spares and donated the airframe to British Aerospace for use as a gate guardian at Warton. XS904 was acquired by the LPG and made the final Lightning military flight to Bruntingthorpe in the hands of Peter Orme, Warton Deputy Chief Test Pilot, on 21st January 1993. Including the six detailed above, a total of seventeen F.6 airframes survive in the UK.

F.Mk.52

XN767, XN770, XN796, XN797, XN729 respectively became 52-655 to 52-659

In response to incursions by North Yemeni ground forces and raids by Egyptian MiGs on Khamis Mushayt AFB and local villages near the Yemeni border, in March 1966 the Saudi Government signed contracts for the urgent delivery of six reconditioned RAF Lightnings (four F.52 and two T.54), six Hunters, two AEI mobile radars and Thunderbird SAM batteries. The four F.2's were recovered from storage (having never entered RAF service) and rushed to the RSAF using RAF Victor tankers by July 1966 to fill the defence gap in an operation code-named Magic Carpet. The fifth F.52 was delivered in May 1967 as an attrition replacement. 52-655 and 656 are preserved but the latter three were all written off in accidents.

Magic Carpet F.52's at Warton, front to rear:
52-655 (XN767) now on display at Dhahran;
52-658 (XN797) crashed Khamis 28-11-68;
56-656 (XN770) now on display at Riyadh;
52-657 (XN796) crashed at Khamis on 20-9-66
[British Aerospace/North West Heritage Group]

T.54 54-650 (XM989) displaying an unusual roundel position which was later changed to the standard position on the nose. The Saudis were also unhappy with the number of fronds on the palm tree and later on a 'bushier' version was used. *[British Aerospace}*

T.Mk.54

XM989 & XM992 respectively became 54-650 & 54-651

Similarly two T.4's were modified and delivered to Saudi Arabia for pilot training, together with two Hunter T.7's. 54-651 was written off after a starter fire but 54-650 survives and is mounted on a plinth at Dhahran.

F.Mk.53

First flight 1.11.66
Saudi: 53-666 to 53-700; Kuwait: 53-412 to 53-423

From September 1968 forty late-development Lightnings (34 F.53 & 6 T.55) were delivered to Saudi Arabia as part of the much larger set of contracts coded Magic Palm, which was signed in May 1966 (although priority work by BAC had begun at the end of 1965). The contract was extended to include 40 Strikemasters, ground radar, airfield construction work and a full servicing and training package provided by Airwork, together with spares and armament. A further export order was obtained in December 1966 for delivery to Kuwait of 12 F.53K and 2 T.55K Lightnings. Again all aircraft were delivered using RAF Victor tankers.

Four brand new F.53's and two T.55's share a hangar at Warton.
The first three are serialled 53-667 (201), 53-670 (205) and 53-668 (202)
[British Aerospace]

The Lightning F.53 was a multi-role version of the long-range RAF F.6. It could be used for ground attack, having hard-points for a pylon under each wing capable of carrying two 1000lb HE retarded bombs or two Matra 155 launchers each holding 18 x 68mm SNEB spin-stabilised rockets. Each overwing pylon could also carry one 1000lb bomb, or twin Matra 100 18-rocket pods as the front portion of 50 gallon fuel tanks; although shown on publicity photographs, the latter never entered production. The under-fuselage pack could be the usual Firestreak or Red Top missile pack, a 44 x 2" SNEB rocket pack or a Vinten 5-camera reconnaissance pack which was available in day or night versions. There were plans for a total of 8 rocket launchers to be carried in addition to the fuselage pack, which could give a theoretical total weapon load of 2 guns and 224 rockets (which were to be fired in waves or salvoes) or 6000lb of bombs. Early trials with the rockets were somewhat fraught, however.

Jimmy Dell: "Initially it looked like someone had just thrown a handful of matches. The rockets were hitting each other and knocking fins off, and when we analysed the films we wondered how we hadn't shot ourselves down. John Nichols (who eventually replaced me as project pilot) was flying the Meteor as photographic chase. I suggested that as long as he flew slightly above and behind me he would be safe, because of gravity drop affecting the rockets. However, when I fired them off he suddenly saw a rocket motor going over his wing! The rockets were programmed to ripple off in fours and we had to change the pattern completely to stop them colliding."

53-666 (modified from XR722, an F.3 held in store) was the first F.53 to fly on 1st November 1966, and on 4th December 1967 53-667 was the first to be delivered to the Royal Saudi Air Force.

53-690 never reached Saudi as it crashed at Stakepool, Pilling (12 miles north of Warton) during its first test flight on 4th September 1968. Pilot John Cockburn was flying on a production clearance flight at 400ft when he 'noticed that the back end was on fire', so he climbed to 4000ft and was heading for Morecambe Bay in order to ditch the jet when he lost elevator control and had to eject, which he did successfully. Mr Bill Lawrenson was checking his newly-laid concrete drive when he looked up to see the Lightning immediately above him. The aircraft plunged vertically through his greenhouse, making a crater 35 feet deep and hurling Mr Lawrenson into the air. He and his wife were taken to hospital suffering from shock and lacerations; his wife had been working in the kitchen and was thrown into a corner when the wall blew in. Obviously believing that Lightning doesn't strike twice, Mr & Mrs Lawrenson still live in the same cottage nearly thirty years later and occasionally turn up small pieces of engine while gardening.

(Above) The scene in Mr & Mrs Lawrenson's back garden near Pilling after 53-690 paid a visit. [Blackpool Gazette & Herald Ltd]

(Left) Mr & Mrs Lawrenson still live in the same cottage 30 years later and have managed to sort out their garden [HT]

The unfortunate aircraft was replaced by 53-700 (later ZF589), the last Lightning built and which was assembled partly from spare components. It was first flown on 29th July 1972, some 2 years after Lightning production officially ended, and at the time of writing lies in a container in Portsmouth.

The F.53 was the only mark of Lightning to see action when in May 1970 an invading Yemeni armoured column was stopped and largely destroyed by successive waves of aircraft using bombs and rockets, the latter proving to be highly accurate and effective weapons. 53-697 was lost to ground fire and crashed on the high plateau upon which Khamis Mushayt airbase is located, but on the Yemeni side of the border (which is only 54 miles from Khamis). However, the pilot *and* the wreckage were quickly spirited away back to Saudi Arabia in a RSAF Hercules.

53-686 lights the reheat for take-off loaded with bombs and under-fuselage SNEB rocket pack.
[Falcon Aviation]

53-686 being armed with rockets at Dhahran in November 1969 before despatch to Khamis Mushayt to suppress the North Yemen rebels. The belly gun pack has the early design muzzles without blast deflectors.
[Falcon Aviation]

Sporting a canopy sun cover and underwing Matra pods at Dhahran with the LCU in November 1969, 53-697 was lost to Yemeni ground fire in May 1970, the remains being recovered by a Saudi C.130.
[Falcon Aviation]

F.53K 53-418 at the Paris Air Show on 6th June 1969 with mock-up twin overwing Matra 100 18-rocket pods married to 50 gallon fuel tanks. Note also the extended under-fuselage pack containing 44 x 2" SNEB rockets. Bombs, guided missiles and overwing tanks are also visible in the ground display.

T.Mk.55

First flight 3.11.66
Saudi: 55-710 to 55-716; Kuwait: 55-410 & 55-411

The Lightning T.55 was the export two-seat operational trainer version of the F.53 and carried the same interceptor armament. It possessed the cambered leading edge and large ventral fuel tank with which RAF Lightning trainers were never blessed, and would have made a superb operational aircraft in Europe. Jimmy Dell observes that the Lightning was a high workload fighter and that although the single-seat pilot could usually cope, in adverse weather conditions "things could get a bit sticky".

The first aircraft , 55-710, was converted from T.5 XS460 (first-flight 2nd February 1966 from Samlesbury to Warton) and flew as a T.55 on 3rd November 1966, but was written off at Warton on 7th March 1967 after a strong cross-wind gust slewed the aircraft into the Rotary Hydraulic Arrestor Gear (RHAG) at the side of the runway, which caused the cockpit to break away from the fuselage. Fortunately, pilots Jimmy Dell and Peter Williams later recovered from their injuries. Most development work was carried out on 55-711 which was built to T.55 production standard and first flew on 29th August 1967. From February 1968 the T.55's were used to train Saudi pilots with 226 OCU at Coltishall, before they were delivered to the Lightning OCU at Dhahran.

Although not a great success in terms of number of airframes sold, the Lightning brought about a great upturn in the fortunes of BAC by facilitating multi-million pound British export sales of defence equipment to Saudi Arabia in particular. Operation Magic Carpet was worth £20M and Magic Palm a further £140M-plus (at 1960's prices) and was followed in the 1980's by the even greater export success of Al Yamamah 1 and 2, which included sales of BAe Hawk and Tornado IDS and ADV variants and which is currently grossing £2 billion per year.

Saudi T.55's 55-711 'A', 714 'D' and 713 'C' whilst based at RAF Coltishall.
[British Aerospace/North West Heritage Group]

In January 1986, as part of an offset deal for Tornadoes, 22 Lightnings in pristine condition were flown back from Saudi Arabia to Warton by RAF pilots and subsequently offered for sale by BAe. They were to be equipped with four Sidewinder missiles, but although Austria and Nigeria showed interest no deal was agreed, and the airframes deteriorated in the salt sea air before being disposed of to museums at very reasonable rates. Sadly, the Saudi authorities have not given permission for the displayed aircraft to be repainted in RSAF markings and so they remain with unrepresentative RAF roundels.

Lightning Missile Installation

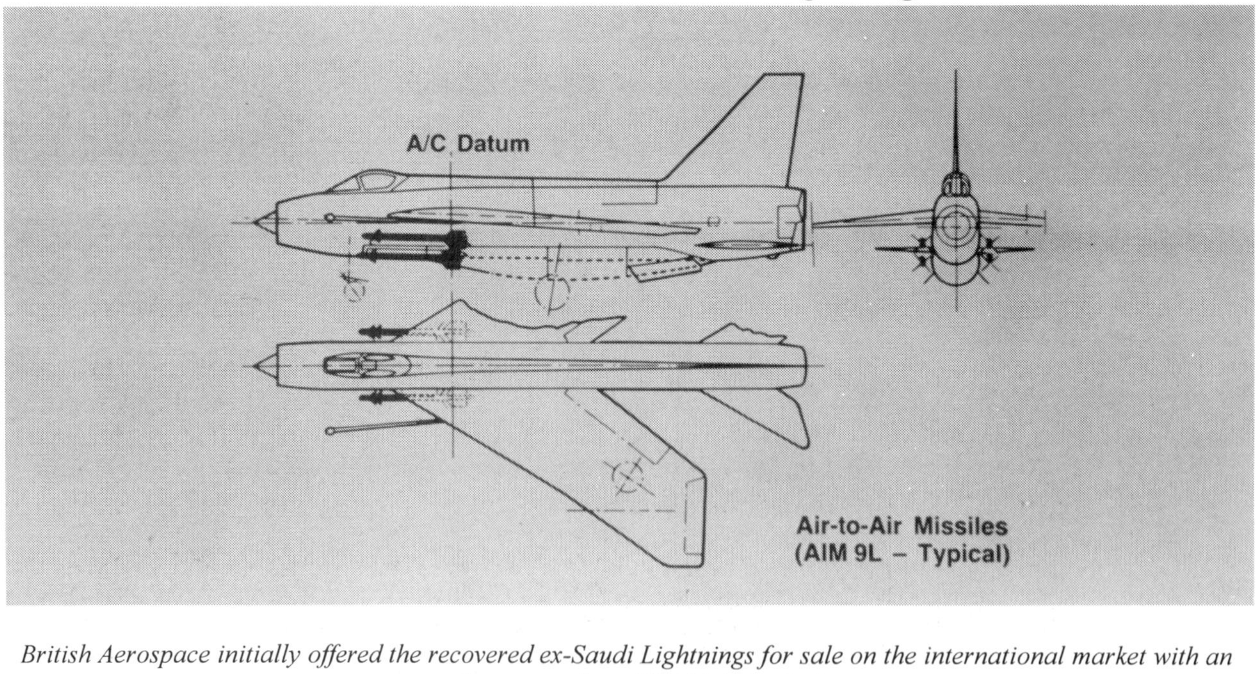

British Aerospace initially offered the recovered ex-Saudi Lightnings for sale on the international market with an upgrade to carry four AIM-9L Sidewinders.
[British Aerospace]

(Top) The F.3A production line looks surprisingly colourful due to the red and blue protective coatings used. The aircraft on the right is XR764 (see also P.32), and at least three T.5's are visible to the rear.
(Centre) Development Batch F.1 XG313 during SNEB rocket firing trials.
(Bottom) F.1A's of the Firebirds line up at Wattisham on 15th May 1963. [All BAe/North West Heritage Group]

Plate 2

(Top) 19 Squadron F.2 XN787 in typical Lightning pose
[Crown Copyright/MOD]

(Below) Lightning F.3's XP751 'B' and XP702 'C' of 74 Squadron cruise through the clouds. 74's dramatic black fin and spine markings were unfortunately later removed and the Squadron badge and fin flash reduced in size on orders from above.
[Crown Copyright/MOD]

(Bottom) An evocative shot of F.2A XN773 in the colourful markings of 92 Squadron, parked in the snow at Gutersloh.

[HT collection]

LIGHTNING SYSTEMS DEVELOPMENT

A few selected Lightning systems are described below: Violet Picture made the transition to production, the others did not.

NAPIER DOUBLE SCORPION ROCKET MOTOR

Jimmy Dell: "The Double Scorpion wasn't flown in a Lightning (despite a well-known retouched photograph) but *was* tested in a Canberra. One problem was that the higher the altitude flown, the more the flames spread and impinged on the rear fuselage. In addition, the groundcrew weren't very happy with handling the rocket fuel and there wasn't any space in the Lightning to carry it. Climb performance was so good anyway that it wasn't really necessary, particularly as the missile armament gave a 15,000ft height advantage. Before the Ministry had discussed it with the test team the requirement was cancelled; the only use for a rocket motor would have been for zoom climbs to intercept at very high altitudes.

Napier retouched a photograph of XA847 (top) to represent a Lightning with a Double Scorpion rocket pack (below). This photograph has sometimes been taken at face value but Lightnings never had this equipment fit.
[British Aerospace/D Napier & Son Ltd]

"We did in fact try a few zoom climbs to see how high the aircraft would go without rocket assistance. Air temperature had to be taken into account for optimum acceleration and a constant energy curve flown. If you rotate into the climb too fast you get a lot of drag which kills off the speed, while if you pull back too slowly you don't use the full potential energy. I did a couple of climbs starting at M2.00+/36,000ft and went over the top at 75,000ft, 180kts/M1.4. The reheats popped out but the engines kept running, which kept all the systems operating. At that height you were virtually a ballistic missile with little or no control and it got to the stage where you had the stick right back and weren't going any higher. Then as the nose came down you started picking up speed again and as you entered denser air and manoeuvrability increased, it was necessary to hold a 3-4G pull-out to avoid going through Mach 2 again as you returned to the start altitude of 36,000ft.

"We passed information on the technique to AFDS and so they did a few interceptions on U.2's, which I gather rather surprised the U.2 pilots. I understand the Lightnings did little more than float by nose-up, but if you got into a suitable position you could theoretically loose off a missile, although at this stage you didn't have any manoeuvrability if you didn't get into the missile launch bracket first time."

DATA-LINK AND AUTO-ATTACK

Datalink was designed to convey target information and ground-computed interception instructions in unjammable digital code which would be converted to visual presentations on the pilot's flight instruments and attack displays; pre-recorded audio instructions were also to be presented. It was also envisaged that in auto-attack mode the aircraft could be steered by autopilot on to the demand heading transmitted from ground control.

Flt Lt Iain Small: "Very early on in the development of the Lightning weapon system, a fully automatic intercept was considered as an option. The system was fitted to one of the prototypes and flown successfully. The auto-attack system used the radar signals from the AI.23B to control the aircraft through the autopilot to make an interception. The radar had to be locked on to the target and would work even if the target was jamming. A modified autopilot control was used. Originally it was designed to be used only above a safe minimum of 30,000ft within the speed range of Mach 0.7 to 1.8, best performance of the system being between 30,000 and 50,000ft with the fighter supersonic. Various safety devices were fitted to ensure that the demands made by the computer would not result in excessive G-forces on the fighter, and the system could be disengaged at any time. The auto-attack system was abandoned in 1965, a bad time for defence projects. It is difficult to imagine the auto-attack system working with hard evading targets and certainly in the 1980's the majority of attacks did not involve using the lock-on function of the radar."

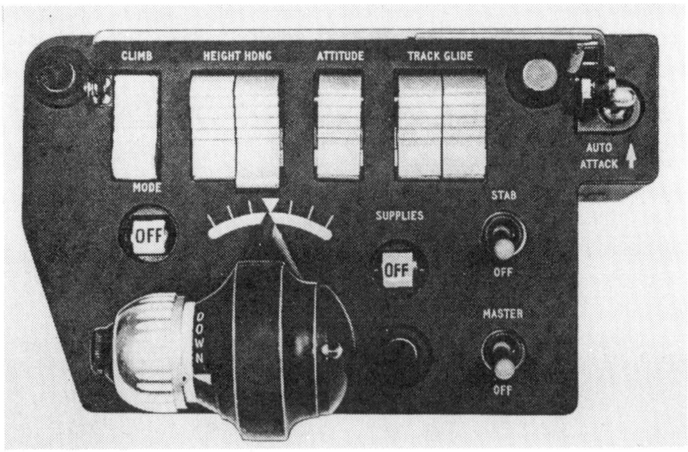

Autopilot control with Auto-attack switch at top right
[British Aerospace]

Jimmy Dell was involved in datalink and auto-attack trials: "The OR.946 system display had small windows which showed command height, speed and heading. On one test with the autopilot engaged a boffin at Farnborough flew me around the Irish Sea by sending commands through Datalink to the OR946/Autopilot system changing height and heading. Command speed changes would come up on the OR946 speed display and the pilot had to manually alter the throttles, but the potential was there for full automation.

"Auto-attack (trialled in XG328) I found fascinating. I did quite a few auto-attacks on Peter Hillwood and the system worked so incredibly smoothly. You had to locate the target on the AI.23, lock on to it, engage autopilot and auto-attack and then the computer took over. When flying *manually* on an attack profile, you had to keep a dot centred in a ring on the screen of the AI.23 to give you the optimum course to steer to achieve missile release. You'd get pretty good at it with practice, but if you were doing supersonic head-on interceptions with a rear-hemisphere missile and got just a little bit late in turning, you were lost—trying to catch up, pulling too hard and killing the speed. The auto-attack system, on the other hand, flew it so smoothly and precisely.

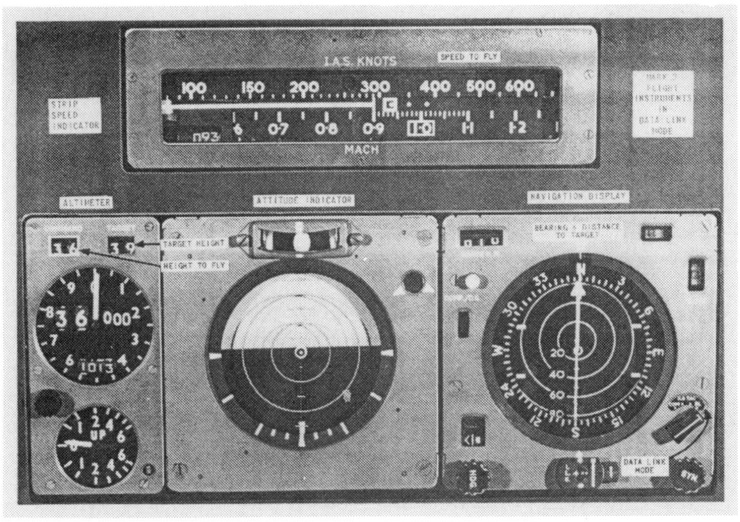

Early F.Mk.3 flight instruments in Data-Link Mode
[British Aerospace]

"It always used to amaze me how soon you had to start turning to get right in behind a target and within firing range. On auto-attack you couldn't even see the target visually when the aircraft began to turn. (If you were flying manually, following the dot and looking out of the cockpit for a contrail, you'd see the target, think he was still a long way off and tend to leave it too late.) However, although a technical success, auto-attack was never fitted to operational aircraft on grounds of cost. The squadron pilots at the time seemed to be coping well without it, the Ground Controllers had been trained up and were making successful interceptions and auto-attack was judged an unnecessary luxury.

"On a related subject, trials were undertaken with Programmed Climb. You could use this with Flight Director selected, in which case you kept the dot centred manually, or you could use Autopilot which would do it automatically. The way it was set up initially, you'd go belting down the runway, select wheels up, switch to programmed climb and autopilot and it would keep you at 250 knots until you reached 0.9M, which resulted in a very steep climb. By the same token, if on a scramble you'd already cranked in your first heading from the Ground Controller, which might be behind you, then it was like an airshow manoeuvre, 250 knots and 60° of bank. I thought that the chaps would never stand for switching to autopilot immediately after wheels up on a dirty night, so we had to back it off somewhat, increase the speed so that the climb angle was shallower and the angle of bank controlled at 45°. Even then it was rather a hairy manoeuvre and if anything went wrong with the system you wouldn't have much time to sort it out. In practice we didn't see the need for the system and so it was dropped."

VIOLET PICTURE (UHF HOMER)

Flt Lt Iain Small: "Violet Picture was a homing system on the Lightning which used the UHF radio receiver and the ILS display format of the OR.946 Navigation Display Unit. It was designed to be used as both an emergency homer on the distress channel and also as an aid to intercepting jamming* hostile aircraft. Two whip aerials located on the fuselage spine provided signals which could be displayed as a left or right indication on the ILS display.

"A hostile aircraft which was jamming the UHF radio would be heard at a much greater distance than the range of the AI.23 radar. By manoeuvring the aircraft so that the 'tramlines' of the ILS localiser display were central, the Lightning would be pointing directly at the target (or, it had to be remembered, possibly directly away from the target if it were behind you!). If the target could be kept central long enough, eventually it would come within range of the radar and so an interception could be made. It has been claimed that by rolling the aircraft on to its side an indication could be gained of whether the target was above or below you!"

***Footnote**: On Practice Interceptions controlled by Ground Control Interception (GCI), Lightning pilots in the target aircraft would jam GCI transmissions to the fighter by (depending on personality), blipping the transmitter, singing, telling very long stories or even screaming on the same R/T frequency.*

SELLING TO SAUDI

JIMMY DELL OBE
Formerly RAF Lightning Project Pilot
and later English Electric Chief Test Pilot

After spending 2 years in Saudi Arabia, Geoffrey Edwards, ex-Group Captain and entrepreneur, informed UK MOD (Ministry of Defence) and English Electric that the Saudi Government was interested in buying 3 or 4 squadrons of Lightnings, a Radar Ground Environment of the latest type, and training facilities. This news was received with a certain amount of incredulity, which I tended to share. At the time I also had some doubts about the potential operating environment as the aircraft had been designed for temperatures up to 35° C whereas the climate of Saudi was closer to 45° C; there had already been problems with overheating as much of the equipment was packed around the engines. However, English Electric's Service Liaison Officer investigated and reported that the Saudis really were interested, so we formed a small team and were despatched to Saudi Arabia. We arrived wearing what we thought were lightweight suits, but which the folks out there regarded as winter wear! The heat was just incredible, 120° F in the shade on the day we arrived, and you could lose a lot of moisture very quickly without realising it due to the dry atmosphere.

Quite honestly we couldn't have sold toffee-apples to children, none of us had had any sales training. When the Saudis said they understood that the Lightning could fly at the speed of sound just above the desert, our Chief Engineer was heard to say "Oh yes, but the cabin conditioning works on a water boiler principle and after so many minutes it would dry up...". Geoffrey Edwards would put his head in his hands in despair. At the time we were mainly in competition with the Americans, who were trying to sell F.5's (the Saudis having declined an offer of F.104's). English Electric had been told by MOD that Saudi Arabia was considered to be within the American sphere of influence and so they should not expect to receive any help with sales promotion. Geoffrey Edwards therefore decided to fly back to try to pull some strings and see, as he put it, "Julian and Alec"—Amery the Minister of Air and Douglas-Home the Prime Minister! We thought, what a name-dropper! Shortly afterwards we were recalled home and on a Monday morning, having been summoned to MOD, we were told: "Gentlemen, at this moment there are four Valiant tankers and 3 Lightning aircraft on their way from Cyprus to Bahrein in the Persian Gulf, for Mr Dell to climb into one at 6 o'clock on Thursday morning to fly to Riyadh and give a demonstration for the King." As this was less than four days hence, panic reigned as we considered all the logistical and operational problems.

As it transpired, I didn't get to Bahrein Airport until 2am on the Thursday morning due to an engine problem on our Comet. Fortunately I was able to get a RAF briefing as 208 Sqn had come up from Aden with their Hunters and the Wing Commander had been told about the proposed Lightning demonstration, although no-one else had as it was considered hush-hush. By the time he had finished the briefing it was about 4am. I was preparing for a 6am take-off when at about 5.15 word came through that the King would be staying at his summer palace in Tiaf and wouldn't return to Riyadh for another ten days. So, with three Lightnings to play with, I had 10 days of flying and swimming, swimming and flying and really enjoyed myself.

Eventually I thought I'd better find the Airport Manager and tell him that I'd be taking off at 6am the next morning in the direction of Saudi Arabia and returning later in the day, which he wasn't very happy about as Saudi Arabia was controlled airspace and required Saudi permission to enter. Bahrein had a very good TACAN but unfortunately the range of the Lightning's equipment was only 200 miles, and so with Riyadh being 250 miles away I would lose the navigation beacon for the final stretch. I worked out I had just enough fuel to be able to do one 360° turn over Riyadh, and then if unable to land I would have to return as there were no diversion airfields.

I took off on schedule the next day in F.2 XN730, just picked out the coast of Saudi and from then on I couldn't see a thing due to extensive sandstorms. I lost TACAN as expected at 200 miles out and made contact with Warton ground crew at Riyadh Airport, but the UHF steering signals were not very good. On ETA I still couldn't see anything and so I began my 360° turn over where I hoped Riyadh would be. Allah must have been with me, as when I got about half-way round the visibility cleared and I saw a railway and a road, of which there weren't many, and then the corner of the airfield itself—it really was a miracle and so I did a quick spiral descent and landed.

A photograph of XN730 taken when in service with 19 Squadron. [MAP]

The day was so hot you couldn't touch the metal of the aircraft without burning yourself. When I did the demonstration it made me realise the effect of that sort of temperature on aircraft performance. I was trying to do a Farnborough-type show but I ended up going miles away from the audience—however they hadn't seen Farnborough so it didn't matter too much. They wanted a supersonic bang, but I was a bit worried about its effect on some of the buildings and so I did my high-speed run, pulled up, peeled over and planted my bang in a dive. I had briefed a chap on the ground to do a commentary, and he told me later that it was a really good double bang which shook the whole hangar beside which everyone was sitting. At first there were looks of horror on the faces of the audience as they thought I'd crashed, until he said, "Gentlemen, you have just heard your first supersonic bang" and their faces lit up with beaming smiles. When I landed they all swarmed enthusiastically over the Lightning.

At last the crowd dwindled away and we refuelled the aircraft. I was just suiting up, when a car screeched to a halt and an Air Force Colonel leapt out to tell me that Prince Sultan, Saudi Minister of Defence, would like me to do a low-level supersonic run through Riyadh. I'd previously had the G90 camera in the nose loaded up with colour film in case anything interesting were to come up and this seemed like an ideal opportunity to use it! There is a nice long boulevard from the airport to almost the centre of Riyadh and so I asked one of our chaps to check that it had no overhead cables. I thought that I'd go through fairly fast but not supersonic due to the risk of property damage, and this is what I did. I think that was the lowest I've flown a Lightning at that sort of speed and to be *invited* to do so was heaven! Just as I pulled up, I thought: 'Dell, you didn't switch the camera on'—which is still a big regret, as it would have been a superb piece of film, and unfortunately I didn't have enough fuel to repeat the run.

For months afterwards I was sent translations of the Arabic newspapers about how many dogs had gone mad and how many miscarriages there had been, but at least no houses fell down. And, to cut a long story very short—we got the contract.

Lightning F.52 52-657 at Warton on 11th June 1966. It was originally F.2 XN746 but saw no RAF service.

LIGHTNING OPERATIONS

MEMORIES OF XR728 —'MY' LIGHTNING

Air Commodore JOHN SPENCER CBE AFC MIPM
Director of Personnel Management (Aircrew), RAF Innsworth
Formerly CO of XI Lightning Squadron & Binbrook Station Commander

When I took command of XI(F) Squadron on 8th April 1982, XR728 had preceded me by a couple of months and was already the bearer of the BA code designating the CO's aircraft, but now I had the enormous pleasure of seeing my name on her side. I had my last trip in XR728, now sporting my initials on her fin, on an Air Combat Training (ACT) sortie on 13th June 1988. In a way I was sorry that my last Lightning flight was not to be in XR728, but that would have been selfish. Far better that she should end her days in pampered retirement at Bruntingthorpe.

During those six years or so we often flew together and my log book records over 100 sorties, more than I flew in any other Mk. 6, varying from the mundane practice interceptions to the vastly more exhilarating ACT and including a fair number of formation displays. Whether as BA or JS I always preferred being in XR728 when we were showing off to the public; after all she was 'my' aircraft.

John Spencer taxying in at Akrotiri in 1982 after leading a Diamond Nine formation in XR728 'BA'. [Mike Hale]

Air Combat Training

Air displays were fun, but ACT was my chief delight and I remember one particular sortie that gave me bitter-sweet pleasure. In early 1988 the Tornado F.3's at Coningsby were working-up prior to taking over from the Lightnings and on a fine April day I was programmed to fly as a singleton against two of them. At this time the F.3 radar was having teething troubles but nevertheless it could 'see' a lot further than could the Lightning's, and the F.3 had a far better overall weapons fit. It was therefore necessary to develop some sneaky tactics to ensure the right result.

Strapping in before the sortie, I mentally rehearsed the opening moves as I waited for the Tornadoes to pass overhead from Coningsby *en route* to the training area. As they did so I started the engines, taxied out, and took off some five minutes later; this way we could set up the combat without wasting fuel.

After checking in with ground radar control, I was told the F.3's were about 50 miles ahead, and after a little positioning they turned in and the game was on. I thought the Tornadoes would probably come in at 5 or 10 thousand feet, so I chose 7,500 feet to give the necessary height separation. Closing in at about 20 miles a minute, it didn't take long for the 50 mile separation to reduce. At 15 miles, 45 seconds away, I selected reheat and pulled into a steep climb, switching off my radar transmitter to avoid them homing in on it, and dropped some chaff hoping to confuse their radar picture. At 25,000 feet I rolled inverted, and levelled off upside down at 30,000 feet. All this had taken about 30 seconds, and as I looked down through the top of the canopy I spotted the F.3's sliding past well below me.

Gentle pressure on the control column put me into a diving attack in their 6 o'clock high position, making it very difficult for them to see me—and they didn't. I slipped in behind the left-hand F.3 and simulated a missile launch right in the heart of the firing envelope, before I was seen by the other one. I transmitted a 'FOX 2 KILL' call, which was relayed to my victim and who was thus out of the fight.

The second Tornado had pulled up in a steep climbing turn to engage me, which suited me fine, as the higher we fought the more apparent the Lightning's superior performance and manoeuvrability would become. With this superiority, it wasn't long before he was nicely in the gunsight at about 300 yards range, and I called a 'FOX 3 KILL' to finish the combat. Feeling highly pleased with myself, we set up for another run......and the same thing happened again! By then of course the Lightning's Achilles Heel, shortage of fuel, asserted itself and so I had to go home, leaving two rather unhappy F.3's in my wake.

XR728 joins the circuit at Binbrook [Peter Talbot]

Of course, since then the F.3 has been modified and improved and the same tactics would not work now; nevertheless, it was good to know that even as its long and distinguished front-line service was coming to an end the Lightning could still hold its own.

Formation displays

Of the numerous displays we did, two stand out vividly. The first was the Queen's Birthday flypast on 11th June 1983. The Lightnings, four of us, were to bring up the rear behind similar formations of Tornadoes, Buccaneers and Jaguars. For fuel reasons we deployed to Wattisham and the idea was for us to rendezvous just south of the airfield with the rest of the 'balbo'.

The rehearsal went without a hitch, and on the day itself the weather over London was fine and clear. However, over East Anglia where we were going to form up into the 16-ship, the weather was awful—low cloud and rain with poor visibility. Well, the Tornadoes and Buccaneers got airborne from Marham and formed up together without problems, before plunging into the clag heading south to pick up the Jaguars and then us; unfortunately, the Jaguars never did find the formation and had to return to Coltishall.

I had set up the Lightnings in a racetrack pattern just north of Wattisham to await the formation. Cloud was forcing us lower and lower, and I wasn't too happy trying to pick up the formation on radar whilst trying to avoid cloud and navigate around the racetrack in lousy visibilty.

The Tornado formation leader kept us well-informed of his position, which he had fixed with his inertial navigation system, but less welcome was the news that he was in cloud at 500 feet. Eventually, I caught what I thought was a glimpse of the formation on radar and turned in hard to take up a position at the rear, but I still couldn't see them visually.

Now, as it happens I enjoy close formation work, especially leading, and this was really why I had insisted that the four of us should have lots and lots of practice, despite the mutterings of those who had not been picked to take part. It certainly paid dividends on the day, as with both skill and faith (probably misplaced), my team hung on for grim death as I gyrated around the Essex countryside trying to find the Tornado/Buccaneer combination.

Eventually the weather cleared somewhat and, more by good luck than good management, we found ourselves in more or less the right position and were able to close up into the correct formation before we started our run-in to the Palace.

So far so good, but all our darting around over Essex had used up our fuel reserves and we were somewhat short when we arrived back over Wattisham, to find that, if anything, the weather had become even worse. Air Traffic had a plan to separate us into two pairs for GCA's, but we didn't have the fuel for this and so I decided on a visual recovery. Luckily my No. 3, Jake Jarron, knew Wattisham well, and so after groping our way to the airfield we broke from close box formation and Jake led us around the circuit, over the church, the garage, the local pub, Farmer Jones' oil-seed rape field and on to the runway. The first beers went down well that evening.

I was reminded of The Queen's Birthday Flypast as the date of the Last, Last Lightning Show drew near. Three days before the event, we had our final rehearsal in the form of a Press Day. The weather was perfect: warm, not too much wind and splendid visibility, and everything went like clockwork. That was on Wednesday 19th August 1987 and over the next two days we watched closely as a series of weather fronts threatened an unpleasant change. However, on the evening of Friday 21st, the met. man assured me that the last front would pass through overnight and the weather would be fine on the day.

I awoke on the 22nd to the pattering of raindrops against the window and the message from the philosophical met. man that a wave had developed on the last front and we were in for a wet day. This time he was right!

As the start of the Flying Display drew near, which my Lightning formation was due to open, the cloud-base measuring equipment went on the blink. It is notoriously difficult to assess cloud-base from the ground and as I needed at least 3,500 feet for the display, I was reluctant to find out the hard way what the actual base was. From the ground it looked about 1500 feet, so I decided that we would open the display with the solo Lightning doing his bad-weather show. Jon Fynes put on his usual splendid performance and reported a cloud-base of some 4,000 feet, which was good enough for me.

Jake Jarron, who had led us around the Wattisham circuit during The Queen's Birthday Flypast, was the display controller and he quickly reorganised the programme to fit us in. With the rain becoming heavier if anything, we manned the eleven Lightnings and prepared to take-off.

All went well at first and the humid air caused nice visible wingtip vortices to mark the aircraft as we took off at ten second intervals along Runway 21. A right turn, 60° bank, holding the speed at 350 knots, brought us back across the runway towards the crowd, before selecting reheat and rotating at 6G into a near vertical climb to 3,000 feet. Rolling over on to my back and pulling down to avoid the cloud, I headed off towards Mablethorpe to form up the Diamond Nine formation.

As I made my way towards the RV, the cloud lowered and the visibility worsened markedly. However, the rest of the chaps followed each other and took up their places in the formation without much problem. Over Mablethorpe we turned left on to a northerly heading, making our way towards the Humber estuary. It was like flying in a goldfish bowl, with sea and sky a uniformly greyish colour and visibility now really poor. I picked out the estuary on my radar as we groped our way up the coast, and highlighted the turn-in point just north of Immingham.

We were then asked to make a delaying orbit as the display before ours was running a bit late. With nine aircraft in close formation in cloud, this was not the most pleasant thing to do, but our previous practices proved their worth and the chaps hung in well and held their positions. I picked up the turn-in point again on radar and with help from Binbrook ATC we turned in, hoping that we were on the right track, as even at 500 feet I still could not see the ground. Gradually during the run-in the weather improved, and suddenly I could see the runway standing out wet and black against the green countryside. Thereafter it was plain sailing, we flew past pulling streamers from the wingtips, and as we turned away the two singletons came roaring through at high speed, almost lost in clouds of vapour.

In retrospect, the Last, Last Lightning Show was the most memorable I flew in XR728, but at the time I was working too hard to really appreciate the experience.

XR728 was always a lady, she never let me down and I never caused her any discomfort either, although some less caring pilots did! She had her own quirks, occasionally at low level the cockpit conditioning would stick on hot and at high level on cold, but the Lightning engendered an affection in its pilots that few other aircraft did, and if the noble lineage of British fighters had to end, then it could have no better conclusion than the Lightning. XR728 is a fine example of that thoroughbred line.

XR728 returns from a sortie with missile pylons empty during a Missile Practice Camp at Valley in September 1987.

[Bob Tuck]

JS reunited with 'JS' : Bruntingthorpe, 23rd June 1991.

[HT]

LIGHTNING PATCHWORK

Squadron Leader P.M. "JIM" JEWELL
Formerly of 56 & 74 Lightning Squadrons

"Just a short article, say 1000-1500 words?"—this was the follow-up to being allowed 20 minutes of 'sitting-in-the-cockpit-wallowing-in-nostalgia' at Bruntingthorpe on Sunday 17th July '94, followed by a dash back to my right-hand seat in the B-17 "Sally B", display and back to Duxford. After 11 years continuous on Lightnings, it was difficult to pick out a one-off incident and remember it in sufficient detail to pad out a couple of pages, so perhaps a small patchwork of my time with the "fleet" would provide an overall series of snapshots.

1. No.3 Lightning Conversion Course (March 1962)

Arrive at RAF Middleton St George (now Teesside Airport), Sunday: unpack, couple of beers, bed. Monday morning: arrival chat, course photo. Three and a half days lectures. Friday morning: quiz, course complete—depart south.

2. No.56 Squadron, RAF Wattisham

a) All the Squadron aircraft were grounded, and in bits, for the hydraulic rework programme and so 'Treble-One' Squadron kindly provided an F.1A for my first trip; no two-seaters in those days, so your first trip was your first solo, but they did give you a 'chase' aircraft to keep you out of trouble! On the afternoon of 22nd May 1962, I strapped into Lightning F.1A XM187 and launched myself in the general direction of the tropopause, followed by Sam in the chase aircraft. Money used to change hands in Crew Room bets: was the 'new boy' going to provide a sonic bang in the climb? With eyes glued to the ASI, the Machmeter needle would slide craftily through 0.98M unnoticed and someone in the coffee bar would collect their winnings. "OK, Jim, let's cancel reheat and return to base." Why? - enjoying self! "Look at your fuel gauges!" Did—went home and logged :40 minutes passenger time. WOW!! What a bird!!

b) In 1963, 56 Squadron took up the Air-to-Air Refuelling trials on the Lightning, refuelling from the single point on Valiant tankers. The early weak rivets around the necks of the probes caused quite a few probe-head snaps initially and resulted in red-faced trips in the Station Flight Chipmunk to Marham to recover the probe-heads, to the accompaniment of much caustic comment from the Tanker crews. With stronger rivets and increasing confidence, the first Lightning squadron AAR detachment to Cyprus took place in February 1964 on Exercise "FORTHRIGHT ONE".

c) Also in 1963, it was announced that 56 Squadron would provide the Fighter Command Aerobatic Team (pre-Red Arrows days) and so the "Firebirds" Team came into being. The highlight of the Season, after five months intensive training, was the Paris Air Show in June 1963.

Firebirds 1963:
(L-R) Brian Allchin, Dick Cloke and Brian Cheater pose on a misty morning at Wattisham with full pressure helmets
[Crown Copyright/MOD]

The Firebirds showing off their paces to pressmen at Wattisham in September 1963
[Top: Peter M Warren; Bottom: Crown Copyright/MOD]

3. No.226 OCU, RAF Coltishall

In April 1963, many Master Pilots (non-commissioned Warrant Officer pilots) were employed in Flight Simulators and doing a first class job. To give a little more power to their elbows as instructors, the policy was to give them flying training on type up to and including first solo; Bill was my student in this scheme. We were having a coffee just after Convex 4, the pre-solo trip, when someone breezed in and said they had heard a rumour that the scheme had now been stopped. A rumour is not an order; within half an hour Bill was briefed, authorised and passing Flight Level 150 in the climb on his solo. The order came in that afternoon and Bill went back to the simulator with a smile. This was an interesting instructional tour as the RAF had started feeding into the Lightning Fleet the first straight-through-from-training pilots. All did well and went on to successful tours on the jet. It began to look like a ground tour coming up for me, but Lady Luck smiled.

4. No.74 Squadron, RAF Leuchars (December 1966)

a) Lady Luck came in the form of an easy decision to forego the "career" path and opt for job/life satisfaction in flying. I joined 74 Squadron (probably the greatest bunch of guys one could hope to work with) as they received the first Lightning F.6's, and with a future of a posting to RAF Tengah in Singapore and the first Air-to-Air Refuelling trip all the way to the Far East.

b) We all looked forward to the arrival of the strange-looking overwing "jugs"—after all, it was only 6 months to the time that we would need them for "Operation Hydraulic" in June 1967.

c) The "jugs" duly arrived and everyone had a trip with the overwing tanks; mine was mid-afternoon in March. At that time of year at Leuchars, night comes early. "Quick as you can, burn off the fuel in the overwings and get back down." Sitting up there in the peace of the evening at 30,000 feet, it seemed a pity to shorten the trip. Oh well—reheat in. BANG/FLASH. "Damn, I'm on fire!" Wrong! The flash of the reheats lighting up had reflected off the inboard walls of the brand-new shiny overwing tanks and focussed straight into the rear-view mirror. Whew! Panic over!

Jim Jewell and XR768 had a close call when a Firestreak went out of control after launch. 768 is shown here with Red Tops and overwing tanks
[Crown Copyright/MOD]

d) Missile Practice Camp, RAF Valley (2nd - 21st April 1967). Here comes the chance to fire off the odd "live" one. After a series of Range Qualification and Photo Chase trips, it was my turn. Target: IR source towed by a Jindivik. Weapon: Firestreak (live). Off I went in Lightning F.6 XR768. The carefully controlled profile of the trip went according to plan, Chris flying Photo Chase and Clive as Standby Shooter. Lock on—Firing Bracket—3-2-1—Fire—a ripping calico noise from the left side of the aircraft and off the Firestreak went. This was when it became a little unfriendly: instead of its control surfaces remaining locked until the missile's required speed had been reached, they unlocked almost immediately. After a few strange aerobatics from left to right across my nose, the "Firework" decided it had had enough and detonated. Apparently it all looked quite spectacular from the Photo Chase aircraft. When asked for a status report by ground control, Chris said: "Target OK. Have to wait until the 'Firer' comes out of the smoke and debris!" XR768 and I emerged unscathed. Chris and I landed back at Valley in a silly cloudbase to have the firing circuits and photo back-up checked out. Clive sensibly diverted to Llanbedr for a beer.

5. Operation "HYDRAULIC" - 74 Squadron AAR from Leuchars to Tengah (June 1967)

a) It sounds fairly simple and straightforward—just flight-refuel "x" number of aircraft from Scotland to Singapore; you have the tankers, pilots and the route. Get on with it.

Just some of the things to think about:

Crew training, flying currency, instrument ratings, briefings.
Engineering/Tech. stores and Publications (spares to be distributed along route).
Engineering personnel for route and destination advance parties.
Squadron Administration (documents, files, publications, office equipment).
Flying Clothing and Safety Equipment (current, route spares and destination).
Vaccinations for all personnel and families.
Logistics of transporting families to Singapore *en masse*.
Handing over of inventories, offices, married quarters, etc.
Boarding schools, selling of cars, storage of furniture.

I did say just *some* of the things to think about!

b) The flying side was relatively trouble-free. My log-book reads:

JUNE 5th	LIGHTNING F.6 XR770	SELF	LEUCHARS—AKROTIRI	5.00
JUNE 8th	LIGHTNING F.6 XR770	SELF	AKROTIRI—MASIRAH	4.45
JUNE 9th	LIGHTNING F.6 XR770	SELF	MASIRAH—GAN	3.30
JUNE 10th	LIGHTNING F.6 XR770	SELF	GAN—TENGAH	4.40

We were supposed to have just the one night-stop in Cyprus at the end of the first leg to Akrotiri. On the way there I managed to pick up the BBC World Service News on the newly-fitted ADF—the Arab-Israeli War was in full swing and we were heading towards it! What timing! Needless to say, the anti-Western Press had a field-day about the British bombers and their fighter escorts coming out to bomb the Arabs. It took a couple of days, and all the Flight Authorisation sheets going to the United Nations in New York, to prove that the "bombers" were tankers, the fighters were unarmed and all we really wanted to do was to get to Singapore.

74 Squadron F.6 as seen through Victor tanker periscope. [Jim Jewell]

On the way out of Gan (Maldive Islands) on the final leg to Tengah, we met up with thick cirrus cloud and embedded cumulo-nimbus tops which gave considerable turbulence. My No.2 and I arrived behind the Victor tanker. As he trailed his hoses for the first top-up, we just sat there and watched the baskets bouncing from four feet above the wing to ten feet below. The phrase: "NO WAY" remained unspoken. We were getting critical on fuel and about to turn back for Gan, when suddenly we were in clear, calm skies—SLURRPP!!!! and all was well.

6. No.74 Squadron, RAF Tengah (Singapore) (June 1967)

A Fire with a Difference - Lightning F.6 XS927 (9th October 1968). I had just carried out a formation take-off (as No.2 to Ian) on Tengah's R/W 18, with a left turn on to North for the climb-out. Abeam the airfield, sliding out into 'Battle' formation for the climb, the RHT 1 (Reheat warning) light illuminated on the Standard Warning Panel, indicating excessive temperature or fire. For those unfamiliar with the Lightning fire system, there is no fire extinguisher for the Reheat (Zone 3) area.

Some weeks earlier, in exactly the same place, an identical situation had occurred with one of the Squadron's newer pilots, with tragic results. The Board of Inquiry established that the fire did not extinguish on shut-down of No.1 Engine and the control runs to the tailplane PFCU (Powered Flying Control Unit) had burned through, causing a pitch-up at low level resulting in an unsuccessful ejection.

I had given this incident much thought and considered that if there had been a fuel leak from the upper (No.2) Engine, it could have found its way down to the No.1 Reheat area immediately below. This would have produced the symptoms, and shutting down the No.1 Engine would have no effect on the fire.

I shut down the "good" (No.2) Engine and switched the refuelling switch to Flight Refuel, depressurising the ventral and flap tanks and stopping any fuel flow from that direction. The RHT 1 Warning Light went out......

I landed—airborne time :05 minutes. My flying gloves were wet.

A "fleet" modification programme of PRC sealant on the deck between the two engines and the provision of drainage holes was immediately introduced.

7. No.226 OCU, RAF Coltishall (June 1969-March 1973)

Having been used to postings every 2-2½ years, a 3 year 9 month tour was almost unbelievable, and magic if the flying was good. This was! :

a) July 1969—July 1970	Lightning F.1A, T.4
b) July 1970—March 1973	Lightning F.1A, F.3, T.4, T.5, F.6
c) August 1971—March 1973	Chipmunk, Spitfires IIa, Vb, XIX, Hurricane IIc (BBMF).

During the Display Season it was like a permanent birthday, with Lightnings from Monday to Friday and Spitfires & Hurricanes at weekends.

The nearest I came to flying XR728 was XR726 at Binbrook on 26 July 1972 during a "Standardisation" visit to 5 Squadron when they lent me '726 for an hour.

XS895 'H' of 74 Squadron taxies in at Tengah, March 1968. [Jim Jewell]

8. What happened after that?

18 years on the VC-10 at Brize Norton seeing the rest of the world, and now an ongoing relationship with B-17G Flying Fortress "Sally B"—more magic.

Footnote: Jim Jewell drew many superb 'Tiger' cartoons for the 74 Squadron Diaries, many of which are reproduced in Bob Cossey's book "Tigers - The Story of No. 74 Squadron, RAF".

ROTATION

PETER GORDON-JOHNSON

Formerly of 5 & 56 Lightning Squadrons
and Deputy Chief Test Pilot at British Aerospace Warton

I well remember the first time I saw a Lightning rotation take-off: it was at Middleton St George in the early 1960's. As a brand new Acting Pilot Officer I was visiting the station during a period of 'holding' prior to going to Church Fenton to start pilot training. Along with a group of young APO's we were standing on the flight-line when this deafening roar occurred and a Mk.1 Lightning thundered down the runway, got airborne, and hung there at what seemed to be just a few inches above the runway. The weather was reasonable with an overcast cloudbase of about 3000 feet. It was only a few moments before the aircraft neared the end of the runway, and then it happened. The Lightning suddenly pitched nose-up, turning a square corner, and powered upwards in a 60-70° climb. It maintained the climb angle until it disappeared into cloud.

It was the sort of spectacle that makes the hair on the back of your neck stand on end. Up to that moment I had only a general idea that I wanted to fly fighters in the RAF but now there was no doubt: I was going to fly the Lightning!

A couple of years later, having experienced some luck, I was at Coltishall starting the conversion phase of the 226 OCU Lightning course. My first experience of the awesome performance of the Lightning was on the first convex (conversion exercise) when my instructor Jim Jewell demonstrated, amongst other things, the take-off and climb performance. We strapped into a very small space that was located way above the ground. After starting-up the engines and completing a few other checks Jim and I taxied out in a Lightning T4 (XM988), lined-up on the runway, and were cleared for take-off. At this stage things started to happen rather quickly; the next 60 seconds passed in a blur. The brakes were released, the throttles slammed all the way forward. A few seconds later I assumed we were airborne as the undercarriage was selected to up; followed by the characteristic loud thump as the nosewheel locked up. Moments later the stick suddenly hit my stomach and the world disappeared. We continued upwards in what felt like a vertical climb until Jim rolled the aircraft inverted to reveal the airfield which was 10,000 feet almost directly below us. I thought that was a good start to a training course, but I wondered if I would ever catch up with the jet.

Five weeks later, on January 13th 1966, I was briefed for convex 22, the last convex with 1 Squadron before moving on to 2 Squadron and the radar and weapons phase. This consisted of a solo general handling and aerobatics flight. During the briefing Jim asked me where I had trained. 'Strubby' I said: I was on one of the handful of courses that did their advanced training on the Meteor. At that time the introduction to service of the Gnat was not going too well. I was not too impressed with a Meteor course because I could foresee postings to Canberras or V-Force—anything but Lightnings. Later on I was very glad to have trained on the Meteor as it was a superb way to learn about asymmetric flying; even better (more demanding) than the Canberra. I also had another superb instructor—Arthur Vine.

Anyway Jim suggested that I do a practice diversion to Strubby, complete a radar approach, clean-up on overshoot, select reheat and do a steep climb away. Needless to say I thought it was a brilliant idea! The brief was to complete the approach to the runway at Strubby, overshoot, accelerate to about 250 knots (this was to be a gentle rotation), pull the stick hard back and climb away steeply; hoping to impress all those watching on the ground.

I walked out to Lightning F.Mk.1A XM188, strapped-in, started-up, and taxied out for take-off. I found my way over to Strubby and positioned for a radar approach. The weather was dreadful with low cloud and rain—the radar approach was necessary just to find the airfield. The cloudbase was low and the visibility such that I could hardly see the dispersal or control-tower. I was not, however, going to miss my first reheat overshoot so, at about 100 feet above the runway I pushed the throttles up to max. dry power, raised the flaps and undercarriage, and selected max. reheat. The aircraft 'took-off' like a scalded cat! I peered through the gloom trying to see the dispersal and ensure I 'rotated' abeam it, but saw very little at all. Oh well, I thought, I'd better just go for it—so I pulled the stick hard back: the world immediately disappeared into total blackness......

XM188 in which Peter Gordon-Johnson rotated vigorously and blacked out. 188 was written off when it taxied into a hangar at Coltishall on 21st June 1968, as described in Roly Elliott's Coltishall article [APN]

I came-to after what must have been just a few seconds later; fortunately with the aircraft climbing very steeply through 3-4000 feet, and still in solid cloud. My heart rate was almost supersonic and it was all I could do to control the aircraft my hands were shaking so much. A few moments later as I got the climb angle under control and pulled the throttles out of reheat to avoid going supersonic over rural Lincolnshire, I realised that air traffic was getting concerned and was calling me. At times like that it was difficult to try and sound as if everything was normal, but fortunately air traffic seemed to think that such a rapid departure was quite normal for a Lightning. One thing was clear, after 22 convexes I was still far from catching up with the Lightning. After the sortie I reflected on the sequence of events and it was clear that in the few seconds I looked for the dispersal, the Lightning had accelerated to 400-450 knots. The only external evidence of my experience was a G-meter tell-tale pegged on or about the G limit of the aircraft. It wasn't surprising that I had blacked-out.

Despite this experience I survived to join my first squadron in the spring of 1966, and halfway through my tour was asked to work up a solo aerobatics sequence in the Lightning Mk.6. This was like manna from heaven, as much of the summer was spent doing one or two practices of the sequence every day plus the occasional display. By this time the Lightning's ability to do spectacular 'rotations' after take-off or on overshoot was legendary and formed an essential part of all displays.

The techniques for rotations after take-off and after overshoot were much the same although the latter was always more spectacular as the fuel weight was less and so the performance was better. In the early days the Lightning Mk.1 started the rotation at airspeeds around 190 knots. Their straight wing leading edges resulted in more buffet during the manoeuvre but strangely less lateral or directional disturbance; they also had the advantage of a better power to weight ratio. The sight of Mike Graydon, and others, snaking up the climb at these low speeds was impressive. The 'conical camber' of the Mk.6 led to a safer rotation if the airspeed were increased by about 30 knots.

The rotation take-off began by lining up at the end of the runway and making sure all the checks were completed. The most important were to ensure the canopy was locked shut, the trims set correctly and the fuel system transferring properly. Unfortunately in the early 1970's a fuel transfer problem on a rotation take-off led to disaster, as the centre of gravity was too far aft and when the pilot initiated the rotation the aircraft pitched over on to its back rather than stabilising in the climb.

When ready to go, the power is increased to check the brakes hold; then released and the throttles set to max. dry power. The jet-pipe nozzles are checked and the throttles slammed to max. reheat. For a moment the thrust stops as the nozzles open to the full reheat position. One to two seconds later the 'hot streak' reheat ignition system lights the reheat fuel and you get an enormous kick in the back as the Avons develop about 15 tons of thrust. There is just time, if you are quick, to check that the engines and reheats have stabilized and check the airspeed indicator is reading about 100-110 knots. A second or two later the nosewheel is raised and aft stick applied to fly off the ground at about 175-185 knots. Once airborne the aft stick pressure is maintained to ensure the aircraft stays airborne (some haven't) and then the undercarriage and flaps selected up without delay. An early undercarriage selection is important as the nosewheel retracts forwards and must be locked up before 250 knots. If it hasn't, air pressure overcomes hydraulic power and the nosewheel can remain stubbornly extended during the climb, much to the embarassment of the pilot and the mirth of the experts on the ground. If all is well the clunk of the noseleg locking-up occurs at 220-230 knots; a good speed in a Mk.6 to rotate. So, gritting one's teeth, the stick is snatched hard back against the stomach: the aircraft pitches rapidly nose-up in moderate airframe buffet, and in about 1 second a pitch attitude of 60° is reached. Probably the most critical part of the manoeuvre is when to check the nose-up pitch rate—about two-thirds into the rotation the stick is slammed almost fully forward to check the upward pitching, and then centralised immediately the climb angle stabilises. The airspeed slowly decays in a heavy Mk.6, so the recovery is started, depending on how brave one is, at about 190 knots and 5000 feet.

The rotation from an overshoot is altogether more fun as the aircraft is lighter and thrust exceeds weight. A normal approach is made and an overshoot started at about 50 feet above the runway at 165-170 knots. The throttles are pushed to max. dry power and the undercarriage and flaps retracted. Acceleration is rapid even in dry power. A moment later, at about 210 knots, max. reheat is selected. The thrust decays as the nozzles go fully open. Less than a second later the stick is snatched fully back to coincide, hopefully, with reheat light-up. The combination of full back stick and a doubling of the thrust kicks the aircraft into a very steep climb. Once stabilised in a 70-80° climb airspeed stays around 200 knots as one powers upwards with the mortals on the ground deafened by the roar of thunder........a great feeling. I liked to maintain the climb to about 1500-2,000 feet and then roll inverted and, pushing the stick on to the forward stop, hold the nose 'up' in an inverted 70-80° climb. By now fuel would be surging through the fuel lines at a rate that would do credit to a fire-hose and, since the normal landing fuel of 800 lbs per side would be approaching fast, it was time to think about putting the jet on the ground.

So at 3,500-4,000 feet the stick is pulled back to complete what is the second half of a loop recovery. Nearing the vertical the throttles are brought out of reheat to control the rapid airspeed increase and a check of speed and height is made to ensure the recovery will be made at 200 feet or higher. On levelling at 200 feet the nose is checked just above the horizon and full roll stick applied for a rapid aileron roll at 370 knots. Wings level, the aircraft is pulled into a 5-6G climb, checking the attitude at 60° nose up. Once again sideways hard with the stick for a 270° roll, followed by a max. G reversal descending back towards 200ft and the display centre. Approaching level flight, max. reheat is again selected, with a brief glance at the fuel gauges, and a 6G pull into the vertical made at 400-450 knots. At 5000ft the throttles are slammed to idle/fast idle and the loop completed, selecting the airbrakes out to try and get the speed under control, with a 90° roll and pull-out on to the downwind leg. Undercarriage and flap are selected down and the final turn made, rolling out on short finals at 180 knots. At about 10 feet above the threshold a slight aft check of the stick is made and the aircraft touches down at 160-165 knots. The nosewheel is lowered on to the runway, the stick pushed fully forward to keep it there, and the brake-chute streamed. Another fun Lightning sortie has ended.

Warning: Do not attempt a rotation unless you are in a Lightning, and then only at a safe height (a height at which you won't strike the tail) !

Peter Gordon-Johnson rotates in XP693 on his last Lightning flight at Warton on Friday 4th September 1992 [HT]

LEUCHARS 'AEROS'

Flight Lieutenant KEVEN MACE

XI Lightning Squadron 1967-69; Based on an article written at the time

"Lightning Mission One-One, two minutes to overhead" . . . the ground controller's voice crackles in the earphones. We are ten miles fron the airfield, cruising at a leisurely five miles per minute, running in for a solo aerobatic display at Leuchars, the No. 1 Fighter Airfield in the U.K. Let's think back to how we came to be here.

"How do you feel about doing aerobatics this year?" It usually starts like that—a casual question from the Boss. The immediate reaction is one of pleasure—and then doubt as to whether you can produce the required standard. Still, the pleasure far outweighs the doubt, so you decide that yes, you would like to perform. From that moment on you are on your own.

Where do you start? After some research into the problem, you find there are a lot of things to think about. The Lightning is a heavy, fast aeroplane—it weighs about the same as two double-decker buses and can travel at 700 mph at sea-level or 1500 mph when six miles high. It cannot fly very slowly; minimum speed when landing is about 200 mph, so we are going to have to fly the aeroplane fast during our display. The faster we go, the more space we are likely to take up . . . but we want to keep near to the crowd during the display, so how do we cope with that? Fuel is going to be a problem—we are going to use it at the rate of 100 gallons a minute, and so we will need to keep a close eye on it and plan our time accordingly.

When we have sorted out the problem of fuel, speed, etc., we now have to work out a series of manoeuvres which will fit into our area and will keep us as close to our spectators as possible. What sort of manoeuvres can we use? Will they fit together so that the speed at the end of one is right for the next one? We know we cannot afford to wait to get the speed right because every 10 seconds we wait takes us a mile further away from the people watching us. These, and many other problems need to be sorted out, but in the end we have what looks (on paper) to be a workable show.

We must next tell the Squadron Commander what we intend to do, and after suitable suggestions for improvements we can go off and try it in the air. First we remain above 7,000 feet, but gradually we are allowed to come lower until finally we arrive at our display height of 500 feet above the ground. Practice follows practice, polishing the rough edges until our display is acceptable to the connoisseurs—our fellow pilots.

Taxying out for display [Terry Senior]

After months of practising, the moment of truth comes: the Battle of Britain display. We've been as far away as Wales, the north of Scotland, and even inside the Arctic Circle to demonstrate this superb aeroplane, but the major show is always the Battle of Britain Commemoration and we will need to put everything we've got into it.

About half an hour before take-off we walk out to the aircraft. Butterflies in the stomach feeling—as usual! Wonder if it shows? Take a quick glance at the sky to check the amount of cloud and its height, the wind direction and the visibility. Looks OK . . . we'll try for the full show. Check around the aircraft, climb up the ladder, strap in tightly so we don't bang our head on the canopy when we are upside-down. Start the Rolls Royce Avon engines, check all twenty-two warning lights are out, call up Air Traffic Control . . . "Lightning One-One Taxi."

We get the clearance and line up on the runway. Open the twin throttles and start the take-off run. Engine temperatures and revs. OK. Push the throttle levers forward through the gate and light the reheats. An extra burst of power surges us forward. In 20 secs we have reached 170 knots and the aircraft lifts smoothly into the air. By the end of the runway we have 280 knots (about 320 mph) and we pull into a steep turn away from the airfield, heading for our holding point on the coast 20 miles to the north-east. We try a quick roll, then hold the aircraft upside-down for a few seconds to check she's flying OK—all seems to be ship-shape, and the time has come to commence our run in. We cross Dundee, heading out to the west of the airfield, bank left to line up with the runway, and head straight in, descending in a shallow dive. We cross the airfield boundary low and fast—400 knots and 250 feet. The front of the Eleven Squadron hangar flashes past and we open up to full dry power, then light the reheats. Twin warning-lights flicker—and then go out—showing us we have two good 'burners'.

Two seconds later we have 450 knots and we pull the aircraft up into a vertical climb, sagging under maximum G. Stop pulling the control column, look sideways to confirm we are vertical, then slam the stick sideways to roll through 180°. Pull on to our back—still with maximum G's. The altimeter shows 6,000 feet—one mile high. Looks OK. Continue to pull and the ground re-appears through the top of the cockpit canopy. We are lined up with the runway, so keep pulling back. The ground rushes up to meet us. But gradually our dive stops and we are back at 500 feet above the runway. Relax the pull. No time to reflect on the last manoeuvre; we're covering the ground too fast for that!

Start to roll left on our slow roll. Kick top rudder to stop the nose dropping below the horizon, keep rolling and we are upside down, hanging in our straps as we push less than weightless to hold level. Continue rolling, the horizon coming back the right way up, feeding in left rudder, and now ten seconds later the roll is complete and we have travelled over a mile to the other side of the airfield.

We roll quickly on to our left side and light the burners again, snatching into a 6G turn. Turn through 90° then roll the wings level and pull the nose up steeply for a wing-over to the right. The nose slides round the horizon, and we crane our head back to look for the airfield. Speed is now low—250 knots, and we dive in a steep turn towards the end of the runway. Check speed increasing—the crowd is down to the left about half a mile away. Speed 350 knots and rising so pull up to the vertical again for a half-horizontal eight, roll and pull through. Height is 4,500 feet this time—just right. Back down, the G force making us weigh about four times as heavy. Closer to the enclosure than the first time, keep the G's on to level out and we are back at 500 feet. The crowd goes by in a blur, the speed slides up to 380 knots, and we thrust the stick to the left and light the burners.

Now we pull the aircraft into a maximum 6G flat turn to the left. Guardbridge and the chimneys flash by under our left wing. The G is now beginning to have an effect—notice we are breathing heavily and feel a trickle of sweat inside the 'bone dome'. The St Andrews road appears directly underneath, showing us we are turning as tightly as possible. Keep the G's on—anti-G suit pressing at our legs and stomach to stop the blood draining away from the head. Don't relax.

The runway is coming back into view and we look out of the top of the canopy. Feels odd, but we are flying on our side so it makes sense. The hangars reappear, then the spectators' enclosure. Keep pulling away from the crowd, then reverse with a 'derry' turn, rolling underneath as we do so, then pull the aircraft round again toward the runway behind us.

Keep it low to illustrate the aircraft's speed, cattle and trees blur underneath as we hold 380 knots. Hold the bank hard on or we will fly too wide.

XR720 on approach [Terry Senior]

Runway beneath us again, roll off the bank fast and pull 6G as we go into a loop. The burners are probably deafening, but in here the only noise is our laboured breathing over the intercom as we pant within the oxygen mask. Climb past the vertical and pull on to our back. Height 4,500 feet, speed dropping off to 250 knots, watch the horizon slide below the nose and keep pulling towards the ground. The aircraft buffets and shakes—we have the maximum lift and air is breaking away from the wings. Our dive looks straight, pointing vertically at the ground just to the right of the spectators. Pull hard out at the bottom, keep the reheats at maximum.

The crowd area is behind us, so roll hard right and turn over the estuary. Roll level and pull the nose up for a wing-over to the left: St Andrews disappears underneath the nose. Roll over left and Tentsmuir Forest appears. Dive down, throttling the engines back to keep our speed down. Bank over the runway, 500 feet, looks about right so roll upside down for an inverted run. Push forward on the control column to hold level. Think how odd the runway looks out of the top of the canopy. Speed 370 knots and the end of the runway approaching.

Push the power up to 100% on both engines. End of runway disappears, so we roll the right way up. Light the reheats and pull into a vertical climb; roll again, the countryside revolves in a blur. Pull on to our back and down again towards the airfield. We keep full power on—the final run is coming up but we must not relax. Twenty seconds to go to the end, pushing the aircraft down to show off the speed. The marker on the airspeed indicator slides relentlessly upwards . . . 500—550—600 knots. We are travelling at 700 mph, or 95% of the speed of sound. Could go faster but not allowed. The whole airfield becomes a blur and we pull up into our finale, the vertical climb. The altimeter shows our height is increasing at a phenomenal rate. . .10,000, 20,000, 30,000 feet in 40 seconds, and we level out, safely out of view of the ground.

Relax at last. We notice how out of breath we are. Check the fuel—just enough left to get us down on the ground, so no problem. Roll over and start a dive back for the airfield. Bank lazily into the circuit and lower the flaps and undercarriage. Over the hedge now and let the aircraft sink on to the runway. Stream the brake parachute at 150 knots and it slows us to a walking pace. Leave the runway and taxi slowly back.

We stop the engines, wave to the wife and look forward to a much-needed cigarette. Feels funny—only six minutes ago we were over Dundee starting our run in, and yet it feels like a lifetime.

XP695 cooling down after a job well done [Terry Senior]

COLTISHALL MEMORIES

Chief Technician ROLY ELLIOTT

Lightning Preservation Group
Formerly of 226 OCU, 60 MU, 5 Squadron & Binbrook LSS & ASF

Two years at Coltishall (1967-69)—what a great time! I finished 3 years of apprenticeship at Halton in October 1967 and arrived at Coltishall extremely wet behind the ears and wondering what these Lightnings were all about. I was soon to find out. On my first day I was walking towards 2 Hangar, blue card in hand, when I peered into 1 Hangar to see a Lightning T.4 sitting on its back end, nose in the air with a large jack poking through the port flap and several Chiefs and Sergeants doing a headless chicken act trying to sort it out. It transpired that the aircraft was being lowered off jacks with no weight in the front and both reheat pipes in: as the troops pulled the main jacks out the aircraft sat down at the back end and to everyone's dismay the port jack modified the flap. Oh dear. That was the start of my learning curve.

The mentor for our group of six new aircraft technicians (multi-trade) was a Chief Technician who looked after us extremely well. For many weeks he would organise us an afternoon off and put us in a classroom where he taught us the aircraft systems one by one and gave us weekly homework to hand in. This was excellent training and got us used to using the aircraft manuals for information. We also worked in all the servicing bays in turn, learning everything from how to strip and rebuild brake units to how to test aircraft instruments. This went on for three months until we were sent out on to the shop floor.

I was sent to the Mk.1's and Mk.4's hangars where I got involved in all sorts of jobs from changing batteries and fire extinguishers to testing the Pilot Attack Sight system. During this spell I went on my first reheat run at night as a spectator. There was no detuner, we did it between two blast walls near the end of the runway, and it was both deafening and spectacular. Seeing those orange/blue diamonds from the reheat pipes at close range with the concrete shaking is something I will never forget. I was also allowed in the cockpit (it was a T.4) for a "clutch", so what a night. I was also learning about the other side of life in "the shed"..... I remember more than once starting a night shift at four in the afternoon and leaving at 8 the following morning, crossing the day shift as they came in. Swings and roundabouts.

I was then moved to "Line Rectification" for the next 18 months, where I thoroughly enjoyed myself, leaving the "sheddies" to concentrate on the really sick ones in the hangar. I learnt to carry out engine runs, do AC/DC checks, change brakes and ventral tanks, as well as doing turnrounds, AF/BF's and 101 other jobs including driving the tractor when no-one was looking. I had some memorable experiences on Line Rects. and here are just a few:

- The early morning shifts in the summer when the "Aeros" pilot would be practising before the CO was up. To see a Lightning inverted at the height of the floodlights for about 10 seconds was one thing, it was quite another when he started to extend the undercarriage.

- The spectacular fireworks display when a pair of pliers went through the No. 2 engine of a T.4 on start-up. An overworked and overtired troop had left them in the intake after a starter change the night before. He couldn't find them and left a note for the morning shift in the Line Rects.' caravan, which unfortunately wasn't used during the surprise exercise that followed the next morning. That added a little more to my learning curve.

- Learning to recognise the sound of Avpin Starter turbine blades whistling past one's ears whilst on a starter crew during which a turbine shed its blades at 44,000 rpm. This was coupled with a world record for the 30 metres as I disappeared behind the nearest Houchin power set. The aircraft's No. 1 engine had refused to start, so the overspeed switch was shorted out to achieve a start, which it did. Unfortunately no-one knew that the control box contacts had welded together, which meant that the starter would continue accelerating until the guy in the cockpit realised and switched the Engine Master Switch 'off'. Of course I wasn't going to hang around to try to tell him, but I think he realised when he saw my smoke.

- The day that one of the aircrew, after landing, shut down the No. 1 engine and taxied in on No. 2 (as usual), not realising that the No. 2 Services hydraulic pump was u/s. That meant that the wheel brake accumulator was the only thing left for his differential braking. Well, the spectacle occurred when the accumulator gave its last gasp just before the last bend into the Line. The aircraft duly went straight on across the grass and buried itself in the Aircrew Office at the end of No. 1 Hangar. The pilot leapt out of the cockpit and raced off down the office roof. As a tribute to Rolls Royce the Avon 210 series engine decided it wasn't going to stop and very efficiently began eating away at the corner of the hangar, the housebricks, window frames, glass, paperwork and In-Tray all getting very well processed at 60% power. The throttle had been jammed in the crash, the problem was how do you stop it?! The electrics had been cut off by the crash switches in the bump and no matter what a few heroes on top did to try and cut the fuel off, it wasn't going to stop. In the end a rather brave person took a panel off the port side, calmly disconnected the throttle at the engine and manually shut the engine down—I raise my hat to him. Well, the aircraft was scrapped, but a couple of years later I bumped into the pilot at Leconfield where he was our post-Major Servicing test pilot. It must have had some effect on him as I've never seen anybody get through a packet of fags more quickly after an air test.

Well, all these happenings and many more certainly jack up the experience level very quickly. Nowhere that I was posted to afterwards had such an effect on me; talk about "read the book, got the T-shirt", etc., I thought I'd seen it all. I hadn't by a long way, but I'll never forget those two years at Coltishall.

XM180 of 145 Sqn, 226 OCU, showing just how exciting life could be at Coltishall as the starboard tyre bursts at the Battle of Britain Display in September 1970 and the aircraft veers off the runway

THE LIGHTNING IN GERMANY

Air Vice Marshal P. S. COLLINS CB AFC BA RAF (Retd.)
Formerly of AFDS, 23, XI Sqns. and CO of 111 Lightning Squadron
& RAF Gutersloh Station Commander 1974-1976

For many Lightning pilots the Mk.2A was the best of the breed. It had the large ventral tank, squared fin and cambered leading-edge of the Mk 6, but it retained the weapons fit of the earlier marks—two Firestreak infra-red homing missiles and two 30mm ADEN cannon in the upper part of the forward fuselage. This combination of guns and missiles together with the increased fuel capacity and aerodynamic imorovements of the later variant made it a delight to fly and a force to be reckoned with in the combat environment of NATO's Central Region in the 1970's.

Only two of the nine Lightning squadrons were equipped with the Mk.2A . These were Nos. 19 and 92, both of which were based at RAF Gutersloh, which I was privileged to command throughout 1975 and 1976. Gutersloh was the most forward of the RAF's airfields in Germany at that time, only 80 miles from the East German border. Consequently, in any circumstances short of war, the "Battle Flight" of two Lightnings which were held continuously at instant readiness near the end of Gutersloh's runway were NATO's first response to any intrusion into Allied airspace in the northern half of Germany. Like the QRA force in the UK these aircraft were fully armed with live missiles and ammunition and the pilots were either in the cockpits or close by in the readiness room, capable of getting airborne well within two minutes of the order to scramble.

19 Squadron F.2A XN777 - Battle Flight
RAF Gütersloh in 1970 [MAP]

There the resemblance to UK QRA ended. Whereas the UK aircraft would climb out over international waters to intercept a distant threat, usually Soviet Naval reconnaissance Bears or Bisons of the Soviet Long Range Air Force, pilots of the Gutersloh Lightnings had constantly to remind themselves that hostile airspace was only minutes' flying time to the east. Their targets, as often as not, would prove to be civil aircraft which had strayed unwittingly into the Buffer Zone, a 30 mile wide strip of sterile airspace intended to prevent unintentional contact with the Soviet combat aircraft based in East Germany. But there was always the possibility that there might be a more hostile encounter. Flying near the border heightened all your senses. You knew that you were being scanned by air defence radars on the East German side. You knew, too, that Soviet fighters were either airborne already or standing by to scramble if you transgressed too far east for their liking. Fortunately, both sides were cautious and professional but we were well aware that NATO aircraft had been shot down in the past after straying over East Germany.

The NATO forces based in Central Europe employed a "belt" system of air defence. The outer defences were provided by surface-to-air missiles with the US Hawk and Nike weapons deployed in forward positions. Behind the SAM belt was a wide fighter engagement zone in which NATO fighters would be free to operate without risking being shot down by their own SAMs. Further back still were the point defences of airfields and other key installations, usually provided by shorter range SAMs, such as Rapier, but in some cases by guns. Ground-based air defence radars scanned the skies constantly by day and night to produce the comprehensive air picture on which operational decisions to launch fighters would be made and by which they would be directed by the interception controllers into the battle.

The second half of the 1970's probably represented the peak build-up of Soviet capability in East Germany and the other Warsaw Pact countries. At Gutersloh, 130 miles north-east of Headquarters RAF Germany and of "The Clutch" airfields (Wildenrath, Bruggen and Laarbruch) where the strike aircraft were based, we were very conscious of what was going on just to the east of us. Each morning our sequence of briefings included an intelligence update on developments in the Group of Soviet Forces Germany (GSFG). We knew that they were steadily replacing their already powerful inventory of combat aircraft with third generation supersonic fighters such as FOXBAT, FLOGGER and FENCER.

These were capable of penetrating NATO territory below the radar cover and attacking their targets without having to expose themselves to our medium level SAMs. At the same time, increasing numbers of heavily armed HIND attack helicopters were being deployed to bases close to the border with West Germany. We knew that our Lightnings and other NATO fighters might be the only real counter to fast low-level air attack, or to mass helicopter thrusts, before the last-ditch point defences could be brought to bear.

F.2 XN787 of 19 Squadron at low level
[Crown Copyright/MOD]

Considerations such as these led us to employ the Lightnings in ways which had never been envisaged when it was introduced into service some 15 years earlier. At that time it was seen as a rapid-climbing interceptor pitted against the high-level threat to the RAF's V-Bomber bases. The theory was that the AI.23 radar would acquire the target and the Pilot Attack System (PAS) would direct the pilot to a missile firing position where it would be unnecessary for him even to see his target before shooting it down. In any event, that simplistic concept had to be modified pretty quickly in the face of sophisticated jamming techniques and the development of radar warning devices, but here in Germany in the mid-1970s we were having to operate in a totally different combat environment. At low-level, overland, the radar could be only an adjunct to the human eye and the primary weapons were often the guns due to the close-range nature of the engagements which ensued. Operational techniques were developed for mounting low-level combat air patrols by pairs of Lightnings deployed in predetermined patterns to cover the predicted attack routes. The Gutersloh pilots became highly skilled in this demanding interception role, flying their aircraft low down over undulating terrain, constantly keeping track of their CAP partner and scanning the horizon for threat aircraft with one eye on the radar and one on the middle distance. Once the threat was sighted, the Lightning's massive acceleration and high manoeuvrability were exploited to close with the target and engage him. These tactics were regularly practised with great success against Phantoms, Jaguars, Harriers and F.104's and there is no doubt in my mind that the Mk.2A Lightning was highly effective in this innovative mode of operation.

At the other end of the operational spectrum was an exercise we conducted once or twice a year with the French and US Air Forces. This was designed to demonstrate the right of the Tri-Partite Powers in occupation of West Berlin to maintain an air access corridor to the city from the west over East Germany. A small number of transport aircraft of each nation would fly the authorised route repeatedly in a designated period, heavily escorted by Mirages, Phantoms and Lightnings. It always passed off uneventfully in my experience, but it was the only authorised incursion by NATO combat aircraft over East Germany and so the adrenalin flowed fairly freely. Gutersloh was usually used as the mounting base for the fighter escort and it provided a useful opportunity for close liaison on the ground and in the air with our French and American colleagues in NATO.

Apart from these rather specialised operational aspects, life on the Lightning squadrons in Germany was not that much different from the UK experience. The annual training syllabus was basically the same, covering all aspects of interception at high, medium and low levels, anti-jamming techniques, air-to-air refuelling and so forth. The squadrons deployed annually to RAF Valley in Anglesey for their live missile firings and to Decimomannu in Sicily for their armament practice camps. One aspect of life became very different in 1976 however, when 19 Squadron moved from their familiar hangar into Hardened Aircraft Shelters (HAS) on the other side of the airfield, followed subsequently by 92 Squadron in another quadrant of the real estate. The RAF Germany airfields became the first RAF bases to experience HAS operations and I believe 19 and 92 were the only Lightning squadrons to do so on a regular basis since Binbrook was never hardened.

XN791 of 92 Squadron prepares for start-up at RAF Gütersloh in July 1972

The brunt of the workload in coping with the new HAS operations fell inevitably on the groundcrew. Accustomed to servicing their aircraft in one hangar and operating them from a line on the concrete outside, each squadron now had to service, arm and launch them from as many as 12 widely-separated shelters. The engineering coordinators could no longer look out of the window and monitor events by personal contact, and ground servicing equipment which could previously move systematically down a line of aircraft was now available to, at most, two Lightnings at a time. What is more, not only were we pioneers in HAS operations but the Lightnings were due to be withdrawn from Germany in early 1977, so no-one was about to give us significantly larger resources of manpower or equipment to cope with these new challenges. Needless to say, these problems and many others were overcome, as they always were, by hard work, ingenuity and the development of practical procedures by our long-suffering and surprisingly cheerful engineering personnel.

In these and in many other aspects of Lightning operations the Germany squadrons found themselves in the forefront of innovation and tactical development. That they succeeded in rising to the wide range of challenges which confronted them is best evidenced by an extract from a letter I received from the then Chief of the Air Staff, Air Chief Marshal Sir Neil Cameron, after a visit to Gutersloh in November 1976. In it he said: "I can sense that the disappearance of the Lightning will be a sad day for you and many of the pilots and groundcrew, but you certainly chose the right way to finish the Lightning era with the best Taceval results for an Air Defence Wing in NATO history." Tactical evaluations of NATO air bases were conducted on a genuine no-notice basis each year by an international team of professional "hard men", reporting to SACEUR, who descended out of the blue and for three days put the base on a war footing and scrutinised the efficiency of every aspect of the station's operational readiness and capability in a realistic and demanding war scenario. Gutersloh and the Mk.2A Lightning squadrons already had a superb record of achievement in NATO Tacevals, but this was their finest result ever. I can think of no better or more relevant tribute to the Lightnings in Germany and to the men and women who flew, serviced and supported them than that.

XN772 of 92 Squadron proving that air combat training was not without its dangers - it was abandoned by its pilot upon entering a spin and crashed at Diepholz on 28 January 1971 *[Günther Kipp]*

THE LIGHTNING AND AIR-TO-AIR REFUELLING

ALAN "PORKY" PAGE
Formerly of 5 & XI Lightning Squadrons and LTF Tactics Instructor

When I was asked to write a chapter for this book the biggest problem was deciding on the subject. Having spent the whole of my operational service at RAF Binbrook, some 7½ years, and flown almost 2000 sorties in the Lightning, it was difficult to know where to start: so, I looked through my log books to try to find a common theme connecting my trips in either XR728 or XS904. On 10th July 1981 I carried out my very first Air-to-Air Refuelling (AAR) sortie in XS904. Now there was a humbling experience, and I even have a picture to remind me of it.

It all started with a one-day course at RAF Marham, home of the Victor K.2 tanker. There were about twenty of us from all types of aircraft on the Receivers' Course. We spent that day learning all about hose drum units (HDU), flow rates and the Victor fuel system; it was truly mesmerising stuff, the highlight of which was lunch! By the end of the day we had learnt that red means stop and green means go. Time to put it into practice.

The AAR conversion exercise (convex) consisted of 6 sorties, 4 by day and 2 by night. When qualified one would be able to refuel unsupervised at any time; possibly in cloud, and with no radio communication (silent). I had heard all the horror stories of the tanking convex phase. The most recent convexee on my Flight had returned with marks on the spine around the Avpin tank and with the ADF aerials broken - this was half way down the fuselage! One of the other pilots brought a Lightning back having attempted (unintentionally) to plug the pitot probe into the refuelling basket—the probe was bent up like a flag pole at the front of the aircraft.

One of my previous instructors had had a major mishap during his AAR conversion. The probe tip, which was riveted on to the end of the refuelling probe, was designed to break off at this weak link if too much lateral force were applied. Many pilots had inadvertently found this out, and either left the probe tip somewhere in the North Sea—or worse, left it in the basket thus making the hose unavailable for use. On this particular occasion the junior pilot managed to defeat the system. Whilst in contact with the tanker he ended up in a very "unusual position". Unfortunately, the probe tip failed to break at the weak link, the emergency brake on the HDU came on and the whole hose broke off, landing in a field somewhere on the North German plain. Understandably, pilots were more than a little apprehensive before AAR Convex 1.

Convex 1 was a solo sortie carried out in a single-seat Lightning, in our case invariably an F.6. Using a single-seater may seem rather peculiar, as the T.5 trainers were also fitted with refuelling probes, but there were good reasons. Firstly, the F.6 had a greater fuel capacity and gave you more time to make contact before you ran out of fuel and had to go home! Secondly, the technique required for the T.5 was different from that required for the single-seaters because of the extra width of the fuselage around the cockpit area. Not only did the refuelling basket react differently when influenced by the airflow around the front of the aircraft, but the pilot's view from the cockpit with the offset seating and separate windscreens was considerably different from that in the F.3 and F.6. Many pilots, myself included, used a slightly different technique when refuelling in the T.5.

On 10th July 1981 I was briefed by Dave Wheeler for a pairs sortie with him leading, to carry out medium level practice intercepts and AAR Convex 1. My aircraft was BD, XS904, and the tanker was to be a Victor K.2. It would all take place on refuelling area 5, the area we usually used at Binbrook, at Flight Level 330 (33,000 feet). There were about 8 designated refuelling areas around the UK where the tankers would maintain a fifteen minute oval pattern (tow-line). Area 5 was controlled by the Boulmer fighter controllers and was about 100 miles north-east of Binbrook.

Dave gave me a very thorough briefing on what to expect, the technique to use and what to do in the case of emergencies whilst in contact. I could use the lines painted under the Victor's wing as a reference, but Dave suggested the edge of the aileron trim tab. (I thought at the time that he was joking but he turned out to be exactly right.) In any event, the aim was to close up slowly from the waiting position, 3 feet behind the hose, concentrating on the chosen references and ignoring the basket. (Its gyrations would only put you off.) The direction indicator perched on top of the left side of the coaming was apparently the best reference of all; just run it up the hose, and as you moved slowly forward you would be in.

As you inched forward, the basket would be affected by the airflow and move out away from the left side of the fuselage, and then move back in to its original position just before you made contact. (On the T.5 the basket moved out in the same way but did not move back in, which required a certain amount of "basket watching" out of the corner of your eye!) If you missed, the supervisor would tell you where and by how much to help you line up for your next attempt. Once in contact, 3 or 4 knots of overtake were maintained to push the hose into the drum until the amber part of the hose was just visible. The green lights would light and fuel was flowing.

As I walked out to the aircraft (having been on XI Sqn only 2 months and with a grand total of 90 hours on the Lightning), I could not help recalling an incident a few years earlier when a Buccaneer had got above the wing of a Victor, where the airflow was very turbulent, and had hit the tanker's tailplane—with fatal results for the Victor crew.

The weather was perfect and 10 minutes after take-off we were talking to the Victor tanker and the fighter controller was vectoring us into position for the rendezvous. No amount of briefing can prepare you for what comes next. As usual, we joined from 2000 feet below and 2 miles astern the tanker, and closed with about 30 knots of overtake to stabilise in formation on his port wing.

The Victor was at about 275 knots Indicated Air Speed (IAS). During flying training you carry out plenty of formation flying but only on similar types of aircraft. Formating on a large aeroplane was an altogether different experience. Dave took up position to supervise me whilst I was cleared astern the port wing hose. Tentatively closing up on the basket, now dwarfed by the huge wing in front of me, I was also increasingly aware of the large fuselage to my right and that vulnerable tailplane almost above my head.

I tried to relax and take my time. As I closed I could feel the distorted airflow around the Victor affecting the controls, and put on half aileron and half rudder trim away from the tanker to balance out the uneven forces. I selected flight refuel and the FR panel lit up; when each light went out the respective tank was full.

I achieved the waiting position about 3 feet behind the basket. This was supposed to be stabilised, but on this sortie my hands were frantically moving on control column and throttles. I desperately tried to relax and not to overcontrol, a lot easier said than done! The lights on the refuelling pod changed from red to amber and I was off towards the basket. Those first few minutes were amazing; I was under the Victor's wing one second and 30 feet astern of the basket the next; up almost level with the wing, then over-controlling down well below the hose. As I closed on the basket it seemed to have a mind of its own and would start dancing around in the disturbed airflow taunting me. I heard it bang on the side of the fuselage and saw it behind the trailing edge of my wing. And then, after I don't know how many attempts of going slowly forward and striking nothing, then sliding urgently backwards to start again, there was this clunk; the hose bounced in the air and the drum inside the pod started to turn, winding in the hose and the lights turned green.

I was in and taking fuel. No I wasn't! Having got so used to sliding back after each successive failure, I responded in the familiar way and closed the throttles—and was out again. After about 3 similar encounters I was actually plugged in and managed to stay there long enough to take some fuel. At least I would be saved the embarrassment of returning to base without having made contact.

After a short time I started to relax a bit and noticed the pain in both hands. I had been squeezing the control column so hard that my right hand was virtually locked solid around it and my left wrist was throbbing from the continual exercise on the throttles!

It took about ten minutes to fill the tanks completely. In a Lightning you never refused fuel as you were perpetually short of gas from the minute you started the engines. The sortie was completed by dry prods on all three hoses, the centre hose being much bigger and heavier than the two wing hoses and requiring a lot more power to push it in. However, once in, the position was well below the Victor and a much more relaxed distance away. The only problem with the centre position was that only one aircraft could use the tanker at a time, so invariably we would tank in pairs using the wing HDU's.

"Porky" Page on AAR Convex 1 in XS904 'BD', 10th July 1981. [Guy Crosby via Porky Page]

On completion we slid astern our respective hoses, moved to the starboard side of the Victor and bid him a fond farewell. (Useful things these tankers!)

I am pleased to report that Convex 2 went considerably better than Convex 1 as I again engaged in a contemporary high speed version of mediaeval jousting in the long suffering XS904 on 17th July 1981. In fact, it really was like riding a bike in that with practice it became second nature. And as a Lightning pilot you got plenty of practice!

It became routine to ask the controller at the start of almost every sortie whether there were any tankers available. When you consider that the Lightning F.6 guzzled one-fifth of its total fuel just to climb to 25,000 feet, it is not difficult to appreciate the reason why. (The F.3 and T.5 would be within 5 minutes of emptying the ventral tank and on to the remaining wing fuel at top of climb.)

The night refuelling trips were another learning experience. Finding the Victor was not a problem using the Lightning's radar and Ground Control Intercept (GCI) radar, especially as it was lit up like a Christmas tree. On a clear night you could see it from miles away. The real judgement came in joining the tanker and sitting on its wing; once close in it was not that different to the day. However, when it was really dark with no moon and thick cloud and maybe a bit turbulent, AAR could become quite difficult, or in the odd case a nightmare.

It was the winter of 1982. I had been Limited Operational for 2 weeks when I spent my second day on Quick Reaction Alert (QRA) in the shed at the end of runway 03 at Binbrook. I was on with Chris Allan ("Spic") and it had been a quiet night with no reported Soviet aircraft activity. We both went to bed just after midnight and I fell asleep about an hour later.

I was woken by the 'phone ringing. Spic answered and then ran out the door in his underwear shouting "Scramble"! I dressed as fast as I could, decided it would be prudent to take thirty seconds for a "visit" and then ran to the aircraft. I had 10 minutes to get airborne from the time of that 'phone call. As I strapped into the cockpit I checked in with Operations who repeated the scramble order. Well, I made it.....just!

Once airborne, in pitch-black moonless conditions, I was handed over to Midland radar and the instruction came to turn on to East and climb to FL 230 (23,000 feet). The tanker was on a 90 degree intercept right to left, twenty degrees right, range 12 miles, at 25,000 feet—radar contact. I turned in behind the tanker about 10,000 feet below him but soon climbed up to his level. Only 15 minutes earlier I had been comfortably tucked up in bed, and here I was stabilised behind the starboard hose, ready to make contact, and we were only just passing the Humber Estuary!

Twenty minutes later I was still stabilised behind the starboard hose attempting to make contact as RAF Leuchars passed a few miles abeam us. I tried all three hoses and must have had an unbelievable number of attempts without making contact. The Victor crew were very understanding, but the more I missed the more tense I became. It was absolute murder. The thought of returning from an operational scramble because I had failed to tank just did not bear contemplation. Not only would I be a laughing stock, but I would probably lose the operational status it had taken so much time and effort to achieve. The more I tried to put this thought out of my mind the more it tormented me, increasing my apprehension.

Well, eventually I did make contact! It was now like Convex 1 all over again. We took up station at 64°N 10°W and were on CAP (Combat Air Patrol) for 2½ hours. This meant I had to plug every 20 minutes to ensure enough fuel to reach Binbrook if for some reason the AAR failed.

As I commenced descent for Binbrook the sun peered over the horizon lighting up the Vale of Pickering and the folding moorland of South Yorkshire. At this point I could actually afford my first smile of the night. No live intercept, the Bears had run for home, but at least I had done what was required.

About a year later twelve of us briefed for a qualification sortie on a USAF KC.135. By this time I had tanked many times and was totally confident in my ability to make contact, usually first time. However, this was a very different "ball game"!

The KC.135 had a boom and the American receivers would hold position while the boom operator would fly the boom into a hole in the top of the fuselage. For probe and drogue operators like us the boom was locked stationary and the hose (which was about 12 feet long) was attached to the boom on a universal joint (UJ) and the basket attached to the hose on a further UJ.

The technique was to stabilise just behind the basket, move gently forward to make contact, then move forward 2 feet, up 2 feet and away 2 feet! This would cause the hose to make a downward "U" shaped loop from the probe to the basket. Unless both UJ's were bent at ninety degrees, fuel would not flow. If you did not get it quite right the hose would do a couple of loops around the probe tip like a skipping rope and take the probe tip off, leaving it in the basket! All 12 pilots had to become qualified over a 2 hour period and I was in the first of 6 pairs. Only one thing was on my mind; not to leave the probe tip in the basket and deny the other 11 guys the chance to qualify.

Luckily I didn't, but this was without doubt the most difficult aircraft to tank from. It required a deft touch, and I found turning at the end of the tow-line particularly difficult. By this time I had developed my own techniques, and like many others had become a "basket watcher"! In fact, on the KC.135 with its shortage of reference points, there was no real alternative.

After the trip I was very relieved to find out that everybody else had found KC.135 tanking as difficult as I had. Sometime later that year the KC.135 supported QRA for a few months—thank goodness I never had to tank from one in cloud at night.

At the time of the Falklands conflict, tankers were in short supply and so a plan was evolved to convert Vulcan bombers into tankers. This was very handy for us at Binbrook as Waddington and the 50 Sqn Vulcan tankers were only 20 miles away. The Vulcan was fitted with the same HDU as the centre drum on the Victor, but right at the tail of the aircraft instead of in the bomb-bay halfway down the fuselage. The result was that the hose flew a lot higher than on the Victor. Furthermore, whereas the Victor's jet pipes were at the trailing edge of the wing, quite a long way away and well above you, the Vulcan's were either side of the hose; it was a face to face confrontation with the hot end of four Olympus engines.

Lightning F.6 XS927 'BH' of XI Squadron refuelling from Vulcan XL445. [Crown Copyright/MOD via Iain Small]

Making contact was quite easy but as you closed up you could feel the vibration of the Vulcan's jet wash on the fin of the Lightning and the rudder pedals would shake around in sympathy. This was later believed to be the cause of loose rivets and some skin damage that started appearing on the fins of our aircraft at about this time.

At this point I have to come clean and admit that I did lose a probe tip whilst refuelling from a Vulcan (or attempting to!). As I plugged into the heavy hose a ripple snaked its way up the hose to the drum and back down to the probe tip within a second or two, and that was it; I now had a drainpipe under my wing with an open hole at the end and the Vulcan had a probe tip stuck in its basket. Time to go home! In fact quite a few pilots managed to join this far from exclusive club. My story was that the hose had obviously not been fully primed with the essential fuel needed to give it stability in flight—and I'm sticking to it!

In 1984 the VC.10 came into service as a tanker. The Victors were nearing the end of their lives, and with tankers in short supply the VC.10 was a welcome addition. The technique was very similar to tanking from a Victor but a little more turbulent on the wing stations. In flight, the wing tips of the VC.10 were well above the top of the fuselage and had a dihedral angle of roughly 10 or 15° to the horizon when viewed from the rear. This had the unwelcome effect of giving a false horizon particularly when tanking at night or in cloud.

The first time I took fuel from a VC.10 was in very hazy conditions at 25,000 feet and about 300 knots indicated. I could not understand why I found myself flying with crossed controls, with aileron applied one way and rudder the other. On glancing at the attitude indicator I found that I was flying with about 15° angle of bank whilst trying to make contact. No wonder I was finding it difficult.

There was quite a lot of turbulence when plugged in to the wing hoses as the VC.10's ailerons were floated up in flight to reduce fatigue. The line-up was exactly down the edge of the aileron and if you moved position left or right just a foot or two the trim required would change considerably as the aerodynamic forces changed. Apart from that it was no real problem, although the first time you sat astern the wing hose, particularly on the left, it was like flying close formation on 2 Conway engines—they were literally alongside your right ear! Glancing up was not a particularly good idea either; that huge tailplane was right above you with a span of almost 34 feet, only one foot less than the entire wingspan of the Lightning! The centre hose was very comfortable, trailing well below the fuselage. Having made contact and achieved the correct position it was hardly necessary to touch the throttles, the hose would just pull you along.

Two 5 Squadron F.6's refuelling from a Vulcan tanker. [Terry Senior]

The Lightning was an excellent aircraft for air to air refuelling, which was just as well really! It was very smooth and responsive in close formation and the engines were superb at height, allowing cold power tanking virtually all the way up. I have taken fuel as low as 5,000 feet and as high as 41,000 feet. It was quite amusing watching certain aircraft struggling to catch the tanker at 25,000 feet, then having to use reheat to make contact. Even at 40,000 feet there was still an adequate reserve of power to enable tanking to be completed without the use of reheat.

Finally, quite a few years after my first faltering attempts at AAR, on 7th January 1988 I was airborne in XS904 over the Irish Sea as a target for radar development trials on the Tornado F.3 from Warton. This time there was a slight difference. Instead of the usual combat fit of 2 missiles and a gun pack as on the standard squadron F.6 (10,000 lbs fuel), XS904 was in ferry fit with the ventral hatch fuel tank and overwing tanks (OWT), giving an additional 4760 lbs fuel capacity.

Doubters at Warton thought that the Lightning lacked the endurance to be a useful target for these trials, but were amazed to find that in this unusual fit it outlasted the Tornado F.3 radar development aircraft (A.03) on task! As a result, the Warton engineers decided to fit the small underwing drop-tanks to A.03 in order to match the Lightning's time on task.

On this particular day there was a possibility of a VC.10 tanker and, much to my amazement, about 1½ hours into the sortie it turned up. I made contact as usual and took fuel. In this configuration the trim change was more noticeable as the weight increased, and obviously more power was required as the overwing tanks started to fill.

After about 10 or 12 minutes, having already taken over 8,000 lbs of fuel, the somewhat bewildered tanker pilot asked me just how much more fuel I was going to take. He had obviously not taken a close look at his rearward facing camera, or seen a Lightning quite like this before. I replied that I would happily take another 3,000 or 4,000 lbs of fuel. There was a stony silence.... I could imagine the crew of the VC.10 feverishly examining their fuel gauges for the cause of this massive fuel leak. Eventually, they admitted to being unfamiliar with the Lightning in this particular fit—they weren't the only ones!

Unfortunately, when the Tornado F.3 went home having completed his task, I was left with over 10,000 lbs of fuel. The maximum landing weight with OWT's fitted equated to no more than 2,600 lbs of fuel so, without going supersonic and without pulling any G, I had to burn it off—needless to say it took a while.

Not bad for a Lightning: 3½ hours on 1 AAR contact!

NON-STOP TO AKROTIRI

BRIAN P. CARROLL

Formerly of 23 & 65 Lightning Squadrons, ex-CFI
and 13 Squadron Royal Saudi Air Force

Lightning Mk 6 XR728

Having been directly involved with the Lightning for some 17 years, I am delighted to be given the opportunity to make my small contribution to this book, adding a little to the history of the Bruntingthorpe collection: specifically XR728. I flew around 26% of the total Lightning fleet and so it was fortunate that XR728, a Mk 6 then belonging to No.23 Squadron, was amongst them.

I first flew XR728 on the 23rd July 1968, the sortie being a high-level practice interception (PI) exercise against Canberra targets; my last flight in the aircraft was on 6th January 1971, a radar mission against unknown targets, ending up with a 2 v 2 combat exercise. During the intervening period I flew XR728 for a total of 57 hrs 40 minutes, or 2.02% of my total Lightning flying, and throughout that time it had a varied and interesting life.

1969 saw XR728 at Akrotiri in Cyprus; GAF Fassberg *(see Footnote)* in Germany; and then on a low level flight via London Airways (probably the only Lightning to do so). In addition, of course, there was lots of routine flying over and around the UK on normal training exercises and on operational QRA interceptions.

In February 1970 I flew 728 to RAF Valley where together we successfully launched one of the earliest Red Top missiles in a head-on attack, scoring a direct hit. The missile was telemetry, not live—that came later, but with another Lightning. In May of the same year we both enjoyed a month of sunshine in Malta.

For the 57 or so hours that I flew XR728 it performed extremely well, only giving me slight cause for alarm on two occasions. The first was on 5th September 1968 when I experienced a No. 1 engine compressor failure, a loud (to say the least) bang followed by a successful single engine recovery to Leuchars. (U/C lowered on the emergency system and no brake 'chute.) The second bit of trouble came on a trip from Leuchars to Akrotiri......

XR728 'D' of 23 Squadron on the line at Leuchars in 1970. [MAP]

NON-STOP TO AKROTIRI - 2,246 nm - time 5 hrs 45 mins

No.23 Squadron (Semper Aggressus) was tasked to detach 12 Lightning Mk 6's from Leuchars to Akrotiri for two months. Myself and another Squadron pilot were tasked with planning the flight, which entailed a detailed look at flight refuelling requirements. This naturally involved a close liaison with the Tanker Force, who did their sums, and ours too. Even so we liked to calculate our own figures just to be safe. Emergency diversions were selected along the route, and diplomatic clearances sought for the various over-flights. The refuelling aspect was quite straightforward, performance charts and slide rules to the fore to work out how much fuel we would need, where to take it on, and of course how many Tankers to meet those requirements. Additionally, the Tanker Force would make their own arrangements as to where their aircraft would land after off-loading several tons of fuel to our hungry birds. (Each of the 12 fighters would use around 9.5 tons of fuel, of which the tankers would provide 6 tons, which made a total transfer for the Squadron of 72 tons or thereabouts.).

Diplomatic Clearances presented no real problems, the negotiating being done by Strike Command HQ. The French (as always) made life difficult by insisting that all flights, including military fast jets, would have to comply with standard airway routes, adding considerably to the distance and time required to transit France. The Italians were usually easier to deal with, normally agreeing that flights could go direct from French airspace south of Nice to overhead Malta. Emergency diversion airfields were also agreed at the same time.

On this particular route our 'bolt holes' were usually Wattisham, Istres, Pisa, Decimomannu, Sigonella and Luqa, although, needless to say, any other port in a storm would do should the need arise. The important thing was that we could always reach one or another of them should an aircraft experience a serious emergency, including a failure of the in-flight refuelling system. The latter was not usually a problem, but Tanker hoses have been known to fail and pilots have occasionally managed to snap a probe head off. This occurs especially in very turbulent conditions when I have seen a centre fuselage hose, which is bigger and heavier than a wing hose, moving through at least 30ft in a vertical arc. Trying to plug into that could get quite exciting. On just a few occasions an aircraft would not accept fuel for other technical reasons and so all of these possibilities had to be catered for.

Returning to diplomatic clearances, these were fine until the actual flights, when 'Roma' Air Traffic Control always tried to change our planned route. Just like the French they would instruct us to fly airways, and being typically British we always refused; well not exactly refused, it just seemed that on such occasions our radios had great difficulty picking up what was being said, so rather than confuse the whole situation we carried on in accordance with our flight plan, informing 'Roma' of our intentions. In any case it would totally upset all of our fuel calculations, and the timing of the RV with our next Tanker. So, expressing our regrets that their transmissions were garbled, we continued on unhindered. This normally resulted in Roma ATC taking the 'huff' and refusing to speak to us again until we were departing their airspace. Still it made for a quiet life. Malta ATC was always very good, their radar making positive contact at quite long range, and they were usually able to pass accurate vectors for the Tanker RV.

Our departure day finally arrived, October 16th 1969, and I had the good fortune to lead the first four-ship off from Leuchars for our 8 weeks in the sun. (The remaining eight aircraft were scheduled to follow on at 30 minutes intervals in two formations of four.) The normal procedure was that the Lightnings would carry out air-to-air refuelling in the Wattisham area to prove the viability of the system, and all being well we would then accompany our particular Tanker across France, allowing the Tankers to navigate the French Airways and do most of the RT. On one of these trips, and it may not have been this one, I was leading a section of four with the usual Tanker escort. We had all taken on fuel as we crossed Norfolk and Suffolk and were now snaking our way over France, the 'French Connection' being due about 100 miles before reaching Nice. All was going nicely as we cruised along at some 420kts, not much to be seen of *terra firma* due to extensive cloud cover, but at around 30,000ft we were enjoying bright sunshine. Yours truly, with nothing better to do, decided to cross-check our position, for although the Tanker was responsible for the navigation we still did our own flight planning. We also liked to know where we were at all times. Having checked and then re-checked on several beacons for a fix, I reached the conclusion that we were some 15 miles off track and heading the wrong way to make good our planned turning point. Dare I suggest to the Tanker Captain that all was not as it should be? Yes I would, and so I called up the Tanker with a polite enquiry as to why we were on the wrong heading for our next turning point and also mentioned that on our present heading we were about to leave the airway!! There was what is commonly known as a pregnant pause: Tanker Captain, "Eagle Red Leader, stand by, I am ***talking*** to the Navigator", even longer pause, "Eagle Red Formation, we are changing course 30° Starboard to make good the next turning point". (Nothing more was said.) Some time later I met up with that Captain in the Mess at RAF Luqa where he bought me a drink (or two), and ***'Navigators'*** never even entered into the conversation.

All four Lightnings successfully topped up prior to reaching Nice, at which point we bade our Tanker *adieu* (he was returning to the UK). I called up Roma Control which brought forth the usual response of, say again your aircraft type, intentions, route, destination, ETA etc., etc. (I often wondered why we filed a flight plan since no one ever seemed to read them.) We were then instructed to change heading in order to join the airways for overflying Italy, at this point my radio failed to understand the fractured English, liberally spattered with excited Italian (one could almost see the frantic waving of arms in the control centre), and we continued on our merry way as planned, i.e. in a straight line for Malta and a meeting with another Tanker.

XR728 'D' peels off followed by XS937 'C', XR747 'K' and XR754 'M' in 1969. Note the large eagle on 'M'. [Crown Copyright/MOD]

We must have been 15 minutes from overhead Luqa when my main flight instruments failed, this coincided with entry into a layer of high-level cloud, with reduced visibility causing the formation of four to close into fighting battle in order to maintain visual contact. I then instructed my No. 3 (the deputy leader) to take over for the RV. This he did for all of two minutes and then declared his radar not working, which would make the Tanker RV somewhat difficult. My Nos. 2 & 4 were Junior Squadron Pilots and not cleared to lead a four-ship, especially in such conditions, so with only my standby instruments working, and no 'on-screen' attitude indication since the MRG was off-line, I took over the lead once again. Malta Radar, as mentioned earlier, were on the ball, and with their assistance and half a radar set a successful RV was achieved, all four refuelled, by which time we were again in clear skies. The forecast was good all the way to Akrotiri so I decided that we would continue in spite of various failures, since repairs at Luqa would present quite a problem.

The remainder of this trip was uneventful; our Luqa Tanker, having refuelled us, landed back on Malta whilst we continued on for Cyprus. The rest of the Squadron all arrived in better shape than my four-ship which had a multitude of snags by the time we had taxied on to the Akrotiri flight line. Five hours and forty-five minutes strapped into a single seater could get a little wearisome, and it was always quite a relief (in more ways than one) to climb down that ladder. XR728 was like every other Lightning, a splendid aircraft—would that I could fly it one more time.

Footnote: *FASSBERG—This was a Berlin Corridor Practice Exercise during which we flew a number of low-level sorties, fitted with overwing tanks, escorting supply planes should the need arise to do the real thing: in the event we were never called upon, but the practice was enjoyable, drifting across the German countryside at speeds somewhat slower than our usual cruise rate, a good speed to view the scenery.*

I LEARNT ABOUT FLYING FROM—XS904

Wing Commander JOHN CLIFFE*

O.C. XI (F) Squadron: Tornado F.3
Formerly of LTF and XI Lightning Squadron

My 'Lightning career' spanned the period April 1974 to July 1982 during which I achieved just over 2000 hours on the machine. Sadly, I then became a two-seat man, running up some 1250 hours in the mighty F.4 and 750 in the Tornado F.3.

My memories of the times I flew XS904 and XR728 are pretty dim now, especially those in XR728, which was a 'low fatigue' target on the LTF when I was a staff weapons instructor. XS904 was one of XI Squadron's aircraft when I first joined the unit in September 1974—on my first tour....and some 17 years later I was lucky enough to be back again as the 'Boss'. In early 1975, about half of XI Squadron deployed to Akrotiri to take over the QRA role from 56 Squadron. Poor old 56 had seen Cyprus torn in two by the war, their families had been evacuated back to the UK, and the crews were confined to camp. I and a chap called Ian Burns were part of the advance party which flew out from Binbrook on a Britannia in the small hours of a Sunday morning in early February. My memories are few, but I do remember that we arrived at Akrotiri in time to be greeted in fine style with a lunch-time 'session' in the Mess bar which introduced me to the delights of Brandy Sours.

* * * * *

Looking through my logbook, one sortie in 904 that stands out was a practice QRA scramble from Akrotiri. We were never told it was a practice scramble until we had identified the target....and this one, from a distance looked remarkably like a Badger. Fortunately, I kept my mouth shut until I was sure what I was intercepting. It was a Nimrod cruising at quite a height, which was no problem for a Lightning. I stabilised alongside and passed details to my GCA controller, who informed me that it was a practice, but that the Nimrod captain had volunteered his services on one condition: that he could get some pictures of me alongside him...upside down.

No problem, I thought, and promptly aileron-rolled inverted. I hope the crew of the Nimrod had high-speed cameras, for I was not with them for much longer. Despite applying full forward stick, a Lightning at 200(ish) knots IAS, at a height approaching 40,000 feet, just did not have enough control authority to maintain level inverted flight. The nose dropped quickly and I lost 10,000 feet in no time. I could almost hear the Nimrod captain laughing his head off, and was convinced that he had set me up....although I had really set myself up. Still, with just 190 hours on the aircraft and still being pretty green, I suppose "I learnt about flying from that!"

Later that day, I bumped into the Nimrod crew in the Mess bar. I told the captain that he had really sucked me in. He replied that his intention had REALLY been to get a picture of a Lightning upside-down alongside his aircraft, and that he had no idea that I would find myself in such an embarrassing situation. He was genuinely concerned, and, as he appeared to be a nice chap, I bought him a drink.

* * * * *

Years later, when I was intercepting a couple of Bear Foxtrots in my fully laden Tornado F.3 which did not want to achieve a stabilised position alongside the targets (and that was the right way up), I couldn't help longing for those halcyon days....short of fuel, with only 2 first generation missiles, no navigation aids worth mentioning, a basic pulse radar, no night vision goggle capability, no head-up display.....but the ability to cruise at 40,000 feet with ease. Hey ho!

**Now Group Captain Plans, HQ Strike Command.*

LIGHTNING FIRE INTEGRITY

Chief Technician ROLY ELLIOTT
Lightning Preservation Group
Formerly of 226 OCU, 60 MU, 5 Squadron & Binbrook LSS & ASF

In the early years of Lightning operation many aircraft were lost due to fires, especially in the reheat pipe bays. The reasons for the fires were many and varied, from the servicing policy, pressure on tradesmen from above to do a 'quick fix', to the unique design of the aircraft giving rise to temperatures of up to 268°C in the reheat pipe bays which were crammed with fuel and hydraulic components. The rate of aircraft loss became so great that a "Fire Integrity Programme" was started. This was first addressed by splitting the aircraft into three Fire Zones: No.1 - The front half of each engine bay; No.2 - The rear half of each engine bay; No.3 - The reheat pipe bays. The design problems were then assessed as follows:

- Rudder and Tailplane control rods passed through all three zones, next to the engines and reheat pipes with no heat shielding. In the event of a fire these would quickly melt and so were changed to stainless steel very early on.
- The fire extinguishers, which were located in the reheat bays, only discharged into the engine bays! In the entire life of the aircraft this was not altered to put fire extinguishant into Zone 3.
- Most fuel and hydraulic components were located in the fire zones, with very few heat shields.
- The integral mainplane fuel tanks passed across the top of the No. 1 engine bay and so any leaks gave immediate problems.
- The No.2 engine bay floor is right above the No.1 intermediate reheat pipe bay, any leaks gave even worse problems.

Action was taken in four main areas which are expanded below:

1. Aircraft Modifications

(a) No.2 Engine Bay Floor

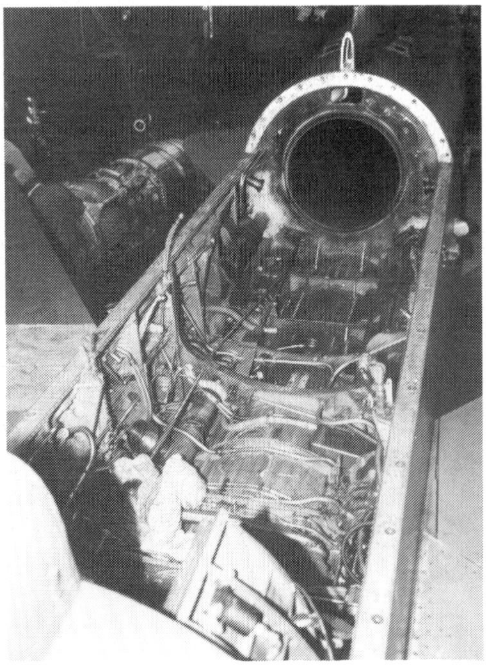

This was a long and arduous modification to seal the No.2 bay floor and prevent leaking fluid getting into the No.1 inter-pipe bay. It was usually done on a major servicing and I was involved in many of these at Leconfield. First the bay was stripped of all components/pipelines. Then it was cleaned and degreased with Methyl Ethyl Ketone. Next, all floor/frame joints and crevices were sealed with 'Viton' and overpainted with epoxy paint. As Viton has to cure under strict conditions, i.e. 18 degrees Centigrade for 72 hours, the bay had to be cocooned. This was achieved by a locally-manufactured 'greenhouse' being placed on top of the bay and all entrances sealed off except the rear of the bays, to which was attached a huge heater/blower on the end of a hoist. A bit Heath Robinson but quite effective.

(b) Drains

All the fire zones were modified to incorporate a drainage system which had to be capable of removing overboard any leaks which might occur. These drain pipes were very fragile and would need constant checking for integrity.

(c) Hydraulic Seals

The temperature in Zone 3 was often above the top-temperature limit of the original hydraulic seals, causing numerous leaks. The bonded seals were therefore changed to higher temperature 'hot' seals, lubricated with silicon grease and located centrally with a PTFE ring. The high pressure unions were then wrapped with Viton tape to turn the high pressure spray from a blown seal into a drip which could then be drained overboard. This was fine in theory, but I wasn't impressed with the research behind it when a wrapped, bonded seal 'went' during a reheat run at Leconfield. The resultant smoke from the burning hydraulic oil, which had totally ignored the wrapping tape, turned a sunny day foggy in an instant and produced a headless chicken display from a few of the crew.

(d) Fuel Seals

The same temperature problem applied here, with the seals plasticising in the heat, torque-loading being lost and leaks occurring. The seals within the fuel couplings in the fire zones were all changed from Nitrile to Fluorel rubber.

(e) Fuel Couplings

Originally most of the fuel couplings in the fire zones were FRS 110 type, which had to be perfectly aligned. Due to the bends in the pipe runs, however, the couplings would move as the pressure increased and so result in leaks. Therefore, a few FRS 380 and 1230K couplings were introduced, as they tolerated being out of alignment, together with extra clipping to stiffen the pipe runs. Great attention was also now paid to torque loading. Whereas previously in the servicing procedure fuel couplings were 'hand tight plus half a turn', this was changed to specific torque loadings for each coupling, with locally manufactured adaptors and expensive torque wrenches appearing on the scene. But the fuel couplings were not finished with yet! Oh no! They now had to be X-rayed after assembly, and the dimensions checked to within thousandths of an inch to ensure perfect alignment before they were passed. All this took weeks and Lord help anyone who stood on a fuel coupling afterwards.

XP706 of XI Squadron in the LESF hangar, January 1983. This aircraft was purchased by the LLPS in 1987
[Stewart Scott]

2. Fire Integrity Checks

It was very quickly decided to do extra checks in the fire zones when the opportunity arose. These were:

(a) An initial examination of the bay, looking for signs of flash fires or leaks.
(b) Leak checks of the fuel, hydraulic and air systems.
(c) A final examination immediately prior to engine/jetpipe fit.

These checks could only be carried out by a "Fire Integrity" qualified SNCO, i.e. someone with a great deal of experience on type. One perennial problem was that it was so easy for the aircraft to become de-modified. One example, for instance, was that if some pipework were removed for a long time and a different tradesman reassembled it, strengthening fairleads and washers could be left off or incorrectly positioned because of inexperience and extremely poor parts catalogues—which often did not show the effects of modifications for years, if at all. With the masses of pipework in the bays minor mistakes could easily arise.

3. Training

The Lightning Servicing School was set up in the early seventies and introduced the 'Safety in Servicing' course which all tradesmen up to the rank of Chief Technician had to take. The course lasted a week and covered all aspects of Fire Integrity on the Lightning. I taught many of these courses and was a great believer in the value of them. Many an old hand would come up to me afterwards and say that he'd learnt a lot from it and wouldn't be taking the aircraft for granted in future. The subjects included Firewire, Fuel Couplings and X-Ray Interpretation, Torque Loading, Main Air System Fitting, Hydraulics and Seals, Safety in Servicing Notes, a Flight Safety film and the obligatory Friday afternoon off.

XS899 'AJ' becomes an 18 ton Christmas card from the Check 2 Team [Terry Senior]

4. Fire Integrity SNCO

Finally, to round it all off, an extra Chief Technician was posted to each squadron to do nothing but Fire Integrity. He would preferably be multi-trade and very experienced in the aircraft. He would co-ordinate all Fire Integrity matters, do as many Fire Integrity checks as possible, report monthly to HQ Strike Command and attend regular meetings there. He was there as a source of information/reference for all tradesmen and was often involved in decision-making in 'grey areas'.

In conclusion, Fire Integrity was a very important part of the Lightning's evolution, as without it many more aircraft would have been lost. It became a way of life, if you didn't treat this aircraft with respect and pay great attention to detail, it would always bite back. Unfortunately it has not been possible here to do justice to the complexity of the subject. I have just scratched the surface in order to give the reader an insight into the problems. No aircraft was designed like it before or since. It is part of the mystique and character of a great aircraft.

QRA-P

Wing Commander PETER HODGSON

O.C. 29(F) Squadron: Tornado F.3;
Formerly of 5, XI and 19 Lightning Squadrons

It has been nearly 14 years since I left the Lightning Force, but I still have a great affection for the aircraft. I spent ten years with the Lightning and flew just short of 2000 hours. My first tour was with 19(F) Squadron flying Mk.2A's in Germany, after which I was posted to RAF Binbrook. Nearly five years with XI Squadron was followed by a further two years with 5 Squadron. It was with XI Squadron that I first flew XS904, on 30th March 1977. I believe it was painted with my name early in 1980 and I then had a long association with the airframe until I left the Squadron in the summer of 1981.

* * * * *

XS901 in the Binbrook QRA Shed
[Iain Small]

My best recollection of XS904 was of an extremely rare QRA scramble on 14th March 1980. I arrived at the Q-shed on time, to find Flt Lt Tony Paxton had taken over the Q1 aircraft and had already been told a scramble was imminent. I accepted XS904 and went to make breakfast. The SOC (Sector Operations Centre) at Neatishead had told us that only one aircraft would be required and that this would be Q1 which would be scrambled in 30 minutes time. Five minutes later as breakfast was sizzling quietly and I was contemplating a pee, we were unexpectedly called to cockpit alert. I expected Tony to be launched as briefed; however, the SOC decided in their wisdom to launch me, because Tony's callsign was the same as the Victor tanker (51, I think) and they wanted to avoid confusion! I was therefore launched without breakfast and urgently requiring a pee!

After 2 hours I was in some discomfort, and never having used a pee-bag in the air had neglected to carry one. After 3 hours I was in agony. Rescue came in the form of control by USAF AWACS aircraft to an intercept of 2 Bear F's in the Faroes Gap. The concentration involved made me forget my acute discomfort and photos were taken, with the usual cheery wave from the tail blisters of the Bear.

MAR	10	LIGHTNING F3	XP 695	SELF	-	V38
MAR	12	LIGHTNING F6	XR 755	SELF	-	T7D
MAR	14	LIGHTNING F6	XS 904	SELF	-	LIVE QRA SCRAMBLE, INTERCEPT 2x BEAR F FAROES GAP.
MAR	15	LIGHTNING F6 -	XS 904	SELF	-	LOSS - LEU
MAR	15	LIGHTNING F6	XS 904	SELF	-	LEU - BK

Totals carried forward

If you refer to my log-book entry, you may wonder why XS904 also features on the following day. On completion of the intercept, my friendly tanker informed me that he had insufficient fuel to return me to Binbrook. I then planned to land at Lossiemouth and refuel. Still oblivious to my bladder problem, I broke into the circuit, only to find the undercarriage would not lower! Emergency drills carried out and gear down, I landed safely, reported to Operations for debrief, went to the Mess and then remembered my bladder! Fifteen minutes later and much relieved, I joined Happy Hour. The following day I returned to Binbrook with gear pinned down, having to land at Leuchars to refuel on the way.

* * * * *

Since leaving the Lightning force, I have had tours on the F.15 Eagle in the USA, the F.4 at Leuchars and the Tornado F.3 (as OC 29(F) Squadron) at Coningsby. The F.15 is a wonderful pilot's aircraft and the F.3 is an excellent all weather weapons system, but the Lightning was more excitement than either, despite its weapons limitations. If the F.3 and the F.15 are the Rolls Royces of our present day, the Lightning was the Lagonda of the 60's and 70's. I applaud your efforts to keep the beast alive.

Bear F: Photos taken by Pete Hodgson on 14 March 1980 from XS904

XS904 as 'D' of 11 Squadron bearing Peter Hodgson's name on APC at Akrotiri in 1980: (above) nearest camera and (below) on approach
[Rolls Royce plc via Peter Hodgson]

SCRAMBLE!

Flight Lieutenant MIKE HALE

56 Squadron Tornado F.3 QFI/CIRE;
Formerly of XI Lightning Squadron

It is a typical January morning at Binbrook, snow laying on the ground, a chill wind blowing and the incessant din of snow-moving vehicles moving around the airfield. I am on QRA with "Big John" Aldington (BJ). Although both BJ and myself are first tourist Lightning Pilots, I am very much the junior; BJ has many previous tours under his belt including Phantoms and Sea Vixens. As the Squadron Diarist, I have just completed my article for the next Station Magazine and, like a fool, given it to BJ to proof-read. As I am being berated about the poor standards of modern education and the lack of literacy amongst junior pilots, the 'phone rings. It is Neatishead, our sector radar centre and scramble authority. A possible scramble is in the offing, two Russian Bear Deltas are transiting around the North Cape. The Bears are flying higher than normal and someone at Command has made the decision to send a Binbrook Lightning instead of the more northerly based Leuchars Phantoms—a very rare flash of common sense. We can expect to be scrambled in about 30 minutes.

The engineers give the aircraft a quick look over to see if all is well, and discover that the liquid oxygen in my aircraft (XS901) had not stabilised after the last servicing, and most of it has evaporated off. BJ, ever the gentleman, offers to swap aircraft and give me his aircraft (XR728) as Q1. We negotiate with Neatishead to have a gentlemanly scramble, with Q1 getting airborne at a prearranged time to meet the Victor Tanker as it heads north past Binbrook. In the event, the tanker is airborne much faster than expected and the alert siren, together with an immediate scramble order, catch us half-way through a last cup of tea. Running in an immersion suit and full flying kit is very difficult; however, when it is to make a scramble time, I don't notice as I turn in my best-ever time for the one hundred yards dash.

XR758 'BF' of XI Squadron in the QRA hangar at Binbrook, 23rd July 1987 [HT]

Up the boarding ladder, jump in, gang-bar up, press starter buttons, then strap-in as the mighty Avons explode into life. As soon as the engines reach idle, the ground crew jump clear and my steed moves forward in special tracks on the Q-shed floor (Lightnings never had the luxury of nose-wheel steering). I can remember the advice of a hairy old Lightning pilot regarding scramble take-off checks: "Canopy down, gyros erect, seat pins out and make sure the runway's clear—the rest is just icing on the cake". I manage these and a few more, as I slip and slide down the recently cleared fast-access track to the runway, trying to balance getting airborne as quickly as possible with ATC, Ops, the Squadron and Neatishead timing me, against the Boss's wrath and my personal embarrassment if I end up parked in a snow drift.

I enter the runway taxiing fast, throttles to full power, a surge of acceleration, a pause to check engine indications, throttles through the gate, a tremendous kick as the reheats light. Almost before I know it I am airborne, gear and flaps selected up, reheat cancelled and radar transmitter on. Five minutes after the scramble bell sounded, I am passing fifteen thousand feet at just below the speed of sound (0.9 Mach) with the mighty Avon thrust-masters purring behind me. With 26,000lbs of cold power available the Lightning has a most impressive rate of unreheated climb. Disappearing below me is Grimsby, full of commuters rushing to work; behind me, as I look down that long graceful wing, is Binbrook, white in its fresh coat of snow and receding rapidly as I climb away. It is one of those winter days when the air is fresh and clear and you can see a hundred miles. At this stage I decide that I am a fraud and that, far from taking the Queen's shilling, I should be paying for the privilege of riding the "stainless steel rocket for taking boy scouts to the moon". Suddenly the radio barks at me. My moment of euphoria is over, Binbrook air traffic are handing me over to Neatishead, there is a job of work to do and to start with, a tanker to be found.

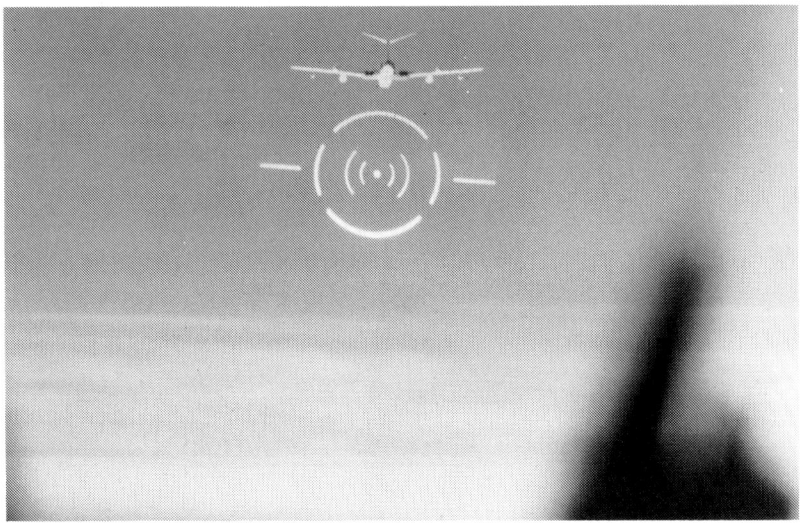

A Victor tanker as seen through the Light Fighter Sight [Mike Hale]

The air is quite turbulent at 30,000ft and as I join the tanker he is climbing through 35,000 for 38,000ft looking for a smoother ride. Smooth air will ease my tanking and help the fatigue life on his airframe. It is important to get plugged in quickly and take on fuel. Not that I need it yet, but once having taken on fuel I am deemed serviceable and so BJ in Q2 can be stood down from cockpit alert. An "Int" up-date on the progress of the Bears is passed by RAF Boulmer. They are heading for the Iceland-Greenland gap instead of the more usual Iceland-Faroes Gap and at almost 40,000ft instead of the more typical 30,000ft. A bit high and north for the Cuba run; perhaps they are having head-wind or turbulence problems like us. Too far north to rush off by myself so I settle down on the tanker's wing for the long transit up to intercept them.

Time to sort out the cockpit after the frantic rush to get airborne. In the Lightning's tiny cockpit I have: three cameras, my own to take pictures for the Squadron diary; my flight commander's, who is about to be posted; the official QRA camera with spare films and batteries; the QRA pack-up full of maps, essential information, code books etc; a packet of sandwiches and various cartons of juice; and, most importantly, a couple of "Bags, Aircrew Relief" for what happens a few hours after drinking umpteen cups of tea and coffee in the Q Shed!

The transit north takes hours but seems to take days. I tank regularly, not only to keep topped up in case I am suddenly required to operate independently, but to keep myself alert. We travel further out into the North Atlantic under the control of Saxa Vord radar station based on the Shetland Islands. As we approach Iceland, control of our little formation is taken by "Pole Star" based on the Faroes. Finally, with about 100 miles to run to the intercept point, I decide to accelerate ahead of the tanker and, keeping enough fuel to divert to Keflavik (Iceland), complete the intercept. Approaching the Bears on a beam aspect the radar starts to pick them up at 60 miles, by 50 miles I have good solid returns, at 20 miles they are visual, and in the last few miles I hear those big Kuznetsov turbo props throbbing away (something I have not experienced before).

As I start about my business of making notes and taking photographs of the rearmost Bear, recording every mark, lump, bump, aerial, window, weapon and weapon station, the Victor Tanker slowly catches up. After a while I notice someone in the large blister window at the rear of the Bear, below its tailplane, beckoning me. As I move closer two crew members behind the window wave enthusiastically, then proceed to show me what appears to be a large bottle of Coca-Cola and a Mayfair magazine; presumably they are proud of their Western possessions. It is then agreed that the Victor will hold station on the rearmost Bear whilst I move forward and down to the leading Bear, who is about 10 miles ahead and a few thousand feet lower. Just as I close to within a mile of the lead Bear I am joined by a pair of American F4s, "Black Knights" based at Keflavik. We exchange a few pleasantries and waves and, as my fuel is getting tight, I decide to leave the lead Bear to them.

As I peel back to find my tanker, he is gone! The trail Bear is there, but the Victor is nowhere in sight and not showing on radar. The Victor, it later transpires, has been working several radio frequencies and received instructions to turn about and proceed to a CAP point; he is now heading in the opposite direction and miles away. My fuel is by now very tight and before long I will have to divert to Keflavik, a nice prospect but another one that will incur the wrath of my Boss and the amusement of my Squadron mates for losing the tanker. Due to the long range, Pole Star are not much help; however, an accurate range and bearing to the tanker come over the airways in broken English with a very strong Russian accent, wishing me *bon voyage* in my English Electric MiG! The information is good, and before long I am plugged in and filling my tanks.

XS904 'BM' of XI Squadron (piloted by Mike Hale) tops up from a Victor K.2, 1984 [via Mike Hale]

The weather back in the UK is worsening across most of the country. As a result the Victor and myself are ordered to return before we reach the designated CAP point and a replacement tanker, already transiting north, is ordered home. On the return trip the turbulence becomes much worse as we pass close to a jet-stream and a weather front. Eventually, we end up having to climb to over 40,000ft, but XR728 is still able to tank without using reheat (about twice the maximum height a Tornado can tank). As we transit down the North Sea, Boulmer passes us numerous weather reports. The weather is deteriorating much faster than forecast and most of the east coast bases are by now 'out', in fog, rain and low cloud. The weather at Binbrook is well below limits and so I am instructed by my Squadron to attempt to get in at Waddington, but to keep enough fuel for a diversion to Valley in Wales or Lossiemouth in Scotland. I do not have enough fuel for this plan and the tanker cannot let me have more; he has just enough to make a couple of stabs at his home base, Marham, and then must divert to Valley. The only stations I can make comfortably are Waddington, where the weather is just within limits, or Leeming, where the weather is a little better. I am all for going to Leeming (as will become clear later) but the Squadron still wants me at Waddington. Fuel will be very tight if I have to try both.

Passing Flamborough Head I leave the tanker, wishing him luck, and descend into the weather front lying over the east coast. I am passed to Eastern Radar and directed by a most pleasant sounding young lady, whose voice conjures up a picture in direct contrast to the gloom I am flying through. As she hands me on to Waddington my main radio fails. No problem—just switch on the standby radio (which works on 243.0, the emergency channel). I am greeted by a deafening shriek—now I have a problem. Taking stock of my situation it is obvious that I cannot get in to Waddington without radio. The TACAN failed earlier in the sortie and I am left with ILS, useful in the final stages of a diversion, and the intercept radar, not much use over land but capable of "painting" the coast and high ground well enough for me to back-plot my position. The Lightning has not yet got a radar altimeter.

I elect to divert to Leeming and set off for the Vale of York. While busy using the radar working out how to get on to Leeming's ILS at a safe height, I pass a gap in the cloud; dropping through, I find the weather to be poor, but fortunately I can see the Great North Road (A1). Although there is a loud howl on the standby radio, I transmit my intentions in the hope that Drayton Centre will pick this up and forewarn Leeming. Fortunately they do and Leeming ATC set all the airfield lights to maximum, so as I follow the A1 northwards I am able to pick out the airfield amongst the mist and land safely.

Naturally, the Squadron do not believe my radios have failed. Reason: my young bride of a few weeks is stationed at Leeming and I am banned from diverting there! Fortuitously the ground crew can confirm that both the radios are out of action, together with the TACAN and, as it transpires, the ILS as well. A broken seal and bent door on the main equipment bay has allowed the entire bay to ice-up, knocking out the various pieces of kit as the long sortie progressed.

However it is not to be quite a happy ending. In training, the Lightning is flown either without missiles or with Dummy or Acquisition missiles, and unlike live missiles the Red Top training rounds do not have wings. When live Red Tops are carried the Lightning becomes more responsive, pitching and turning better when manoeuvred aggressively, because the large Red Top wings act like 8 canards and move the centre of lift forward. Normally, live rounds are only carried on QRA or Missile Practice Camp sorties which consist of fairly docile profiles. However, the following day when I leave Leeming to return to Binbrook, I decide to fly a classic "Lightning rotation" takeoff. XR728 still has her missiles on (made safe) and of course pitches up into the vertical more quickly than I anticipate, before climbing away in fine fashion. Although far from overstressing XR728, I use more "airframe fatigue" than is allowed on a straight transit sortie (Mark 6 fatigue is saved for good deals like air combat sorties).

So it came to pass that I did finally incur the wrath of my boss (Wg Cdr John Spencer) and was awarded a weekend as Station Duty Officer for my troubles—some you win, some you don't. Needless to say, the boss did not bear a grudge and a few years later he allowed me to fly my wife in a T-Bird Lightning!

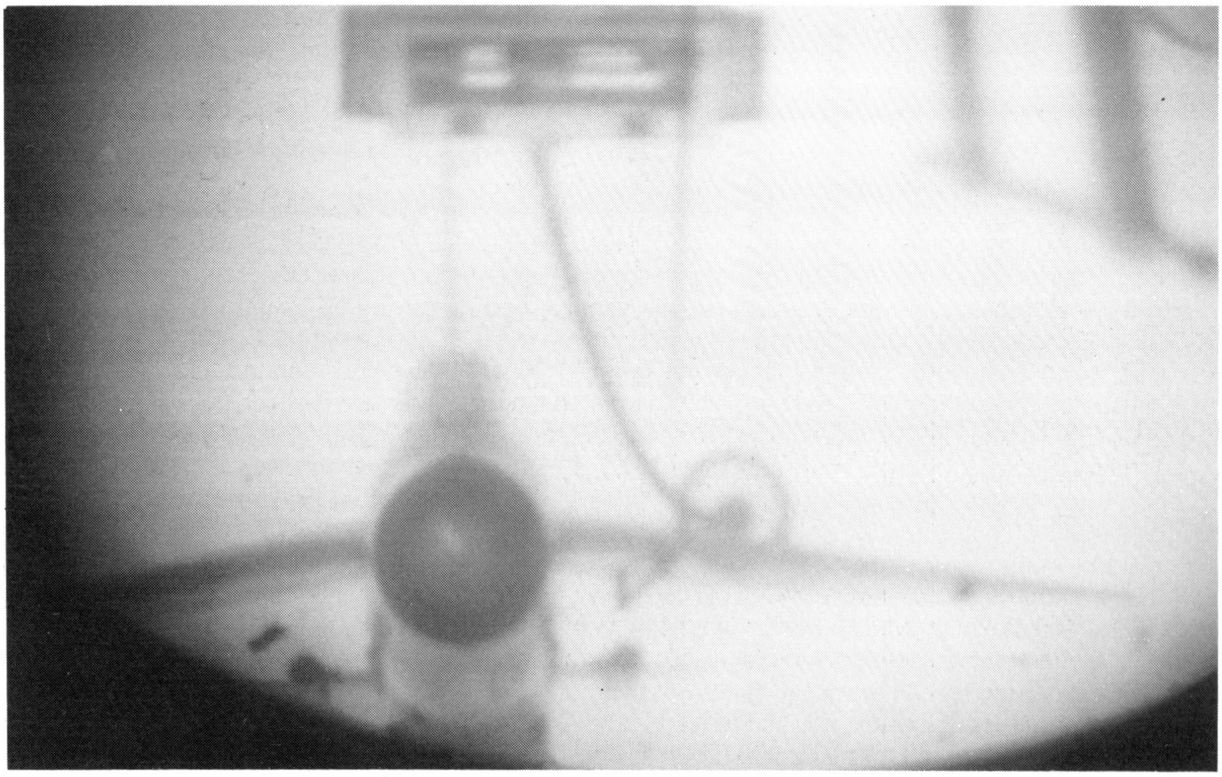

Operation Fan-Angle Charabanc, Winter 1985, Keflavik: XS904 'BM' of XI Squadron, flown by Mike Hale, seen through the 'Boggle' Scope of a Vulcan tanker. [via Mike Hale]

LIGHTNING AIR COMBAT

Flight Lieutenant IAIN SMALL
Formerly of XI Lightning Squadron

The original concept of the Lightning was for a very quick reaction interceptor that excelled at high altitude and high supersonic speed. The envisaged targets in the early days were the high altitude and sometimes supersonic Soviet bombers. Well, the Lightning certainly did excel in that environment but with the introduction of the F.2 to Germany in 1964, and later the F.2A, the aircraft was to encounter new threats. Gütersloh, being very close to the East German border, meant that quick reaction and high speed were still important, but in the event of hostilities the targets would not be the high altitude bomber but the fighter bomber with a MiG fighter escort. The Lightning became a very potent dog fighter and the second-to-none reputation of the Gütersloh squadrons is legend.

With the realisation of a possible fighter threat to the UK, with longer range Soviet fighters with tanker support and also more recently aircraft carriers, the UK Squadrons also carried out Air Combat Training. The squadrons could also be deployed world-wide where they might be up against fighters.

Air Combat Training on XI Squadron in the 1980's initially took place mostly in the Lightning F.3. The F.6 aircraft by this time were a good way through their fatigue life and the remaining fatigue had to be looked after very carefully. The Lightning F.3 had a better power to weight ratio than the F.6 as it was considerably lighter and the handling was slightly different due to the different wing plan form. The down side of the Mk.3 was a lot less fuel. With about 5000lbs of combat fuel available after take off, and with a burn rate of 1000lbs per minute, it was not unknown for a total flight time of 10 minutes which included only 5 minutes of combat training 10 miles away from base. It doesn't sound much but it was a very exciting 10 minutes and a lot could be learnt. Later when some of the Mk.6s were given modifications to extend their fatigue life, we could stay in combat much longer.

XP741 of the LTF being flown by Iain Small [via Iain Small]

The incredible aerodynamic performance and excellent handling qualities of the Lightning meant that it was a very good performer in the air combat arena. It would always do extremely well against all RAF adversaries and indeed most of NATO. Only when up against aircraft such as the F.15 and F.16 would we lose the performance advantage—that's where tactics came into play.

Generally in air combat the maximum speeds achieved were about M1.1 to M1.2 but for most of the fight it would be much less than that. With its superb power to weight ratio the Lightning would accelerate very quickly at around 25 knots per second at low level. In a turning fight a very respectable turn rate of 12° per second could be sustained in maximum reheat, pulling 6G at around 450 to 480 knots. Slightly higher instantaneous rates of around 14° could be achieved at 400 knots, this being very useful to avoid a missile for example.

One particularly impressive aspect of the Lightning was its performance in the vertical plane. The aircraft had an initial rate of climb of around 50,000 feet per minute and in a turning fight at low level could sustain a 5G turn and still climb at 18,000 feet per minute. Clearly if other aircraft were struggling to sustain 5-6G in a level turn at low level, the Lightning would be able to climb as well as turn thus allowing more freedom of manoeuvre.

The problem area for the Lightning in combat was without doubt its weapons system and radar. Optimised for medium to high altitude intercepts against bomber targets the system left a certain amount to be desired. Against a fighter the AI.23, being a pulse radar, had a very limited range of about 15-20 miles looking up, and looking down suffered from ground clutter which reduced the pick-up range considerably. In its favour was the fact that its high sweep rate and ease of handling were ideal for air combat where the targets would be manoeuvring wildly both in direction and altitude. In air combat the radar would be used on the run-in to the fight to try and get the target visual. Once visual the radar would largely be redundant unless the target was lost, possibly flying through cloud or towards the sun. Without a Head Up Display giving target information, the most difficult aspect of the fight would sometimes be trying to convert a blip on the radar to a visual target outside.

Iain Small firing a Red Top at a towed flare (top left) from an F6 at 15,000 ft. Horizon is visible at bottom left [via Iain Small]

The missiles, both Firestreak and Red Top, had to be fired within a narrow cone of the target's jet pipe and could be used if the targets were not evading too hard. Once in a hard turning fight the guns were the primary weapon.

Lightning tactics, on XI Squadron in the 1980's, involved primarily a pair of aircraft and most of our training involved fighting against dissimilar types. Typically a fight would be a 2 v 2 but other options such as 2 v 1, 1 v 2 and 2 v 4 would also be practised. Transits to the combat patrol area would be at a fuel-economical 360 knots. Once on the combat patrol 400 knots would provide reasonable economy but with good instantaneous turning capability if bounced. When engaged in a combat, ideally the speed would be about 450-480 knots for best turning performance. This is a good speed to fight in a high threat environment with lots of targets, as it leaves many options available. Once engaged in the proverbial 'knife fight in a telephone box' against a single target at slow speed it is very easy to be shot by another unseen hostile aircraft. In a 2 v 2 the ideal situation would be for a pair of Lightnings to kill one target, giving each other support, and then go on to kill the other. More often the fight would degenerate into two 1 v 1 combats, probably close enough for an element of support to be given.

In February 1985, Sea Harriers from Yeovilton and Hawks from Chivenor had been invited up to Binbrook for a week of combat and tactics discussions, normally held in the Blacksmith's Arms at Rothwell! On 6th February I flew XS904 on a 1 v 2 v 2 combat against Sea Harriers and Hawks; a very hostile environment. The Hawk is a very small agile fighter with good missiles and is very difficult to pick up. The Sea Harrier is not quite so agile but has a very good weapons system. The plan was to fly the combat over the North Sea just to the south of Flamborough Head with the help of one of the east coast Air Defence Radar sites.

A Sea Harrier FRS.1 on the line at Binbrook [Iain Small]

The Lightning Wing at Binbrook had a very close liaison with the controllers (I actually married one!!) who were very adept at the type of control we needed. The fight started with the three elements split by 20 miles on each corner of an imaginary triangle. Running in from my start point just subsonic at 0.95M the controller started giving me information on the other two formations. I decided to aim at the Hawk formation. As they were without a radar, I had a chance of a quick kill without being seen. If we got tangled up with the Hawks for any length of time the Sea Harriers would come in and kill everyone.

As the controller called the Hawks at 8 miles and at slightly higher level, I picked them up on the radar. At this stage the Red Tops had been armed but the guns were selected for firing. The selector to change from missiles to guns was on the right-hand console. This was not a terrific place for it especially when pulling 6G with your right hand on the control column. By keeping guns selected, they would always be live for a guns kill or a lucky snap shot, and if a missile shot did present itself, it was then not a problem to make the selection. Approaching the merge I managed to get visual with one of the Hawks, but not the other. I had the throttles back at idle to give a low heat picture to their Sidewinder 9L missiles. As it would be suicidal to turn in on one Hawk I selected full reheat and accelerated straight through. The controller called that the Hawks had not turned in on me but were heading for the Sea Harriers.

When 10 miles clear, I turned back in and was immediately given a vector by the controller. The Hawks and Sea Harriers were now fighting each other and had split into two 1 v 1 fights separated by 5 miles or so. I ran in towards one of the fights and picked them both up at about 5 miles. A change to missiles, with the sight now on the jet pipe of the Hawk, the Red Top acquired. I gave the call 'Fox 2, Hawk left-hand turn at 8000'. The Sea Harrier quickly turned into me to try for a shot. I again put the throttles to idle to counter the Sidewinder and passed him head to head very close. Another maximum reheat acceleration took me clear of the Sea Harrier and towards the south west and home, as usual with no spare fuel remaining. A 30 minute flight, 10 minutes of combat, with 1 kill achieved and no claim against. Not a bad result, but what it could have been with a pair of Sidewinder 9Ls!!

G.90 gun camera film taken by Mike Hale in XS904 against (top to bottom): F5E Aggressor, F4 and F15. Lightning pitot visible at bottom. [Mike Hale]

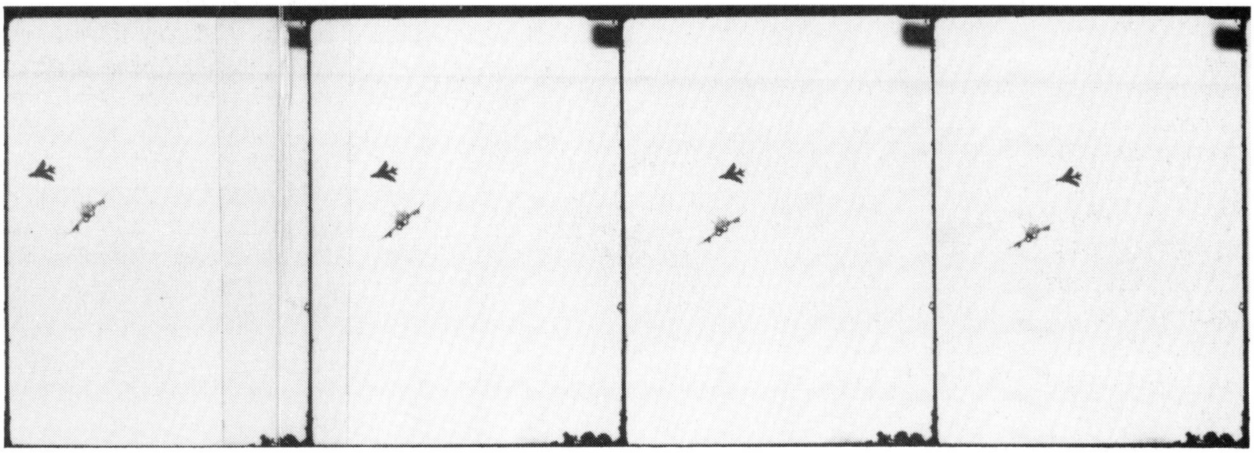

Lightning chasing Phantom chasing Lightning. Taken by Mike Hale flying XS904; the Phantom was flown by Mark Hanna. [Mike Hale]

TREBLE-ONE AT WATTISHAM

Chief Technician PETE O'CALLAGHAN
Lightning Preservation Group
Formerly of 111and 29 Lightning Squadrons
& TFF, Wattisham; 5 Squadron and ASF Binbrook

During the early seventies I was a mechanic at Wattisham servicing Lightnings on 111 Squadron (Treble-One being popularly known as The Tremblers). It was a good squadron, members worked hard and played hard. In 1974 the Squadron obtained three Mk. 6 Lightnings to add to its fleet of Mk. 3 and Mk.5 aircraft, and they were given the tail letters X, Y and Z. During one exercise the Squadron Commander felt that he needed to give a pep-talk and said that the Squadron needed more zest in their approach to work. That night, XS895 'Z' acquired a few more letters.....on the line next morning the Squadron found ZEST written in black tape down its fin.

On detachment in Malta our team had an engine change to do. We cracked on with the job and got the engine out, prepped the new one to go in and put the starter dome inside the aircraft intake ready for fitting, as the engine won't fit if it's already attached. We fitted the new engine, getting on with it quickly as we were wasting valuable drinking time. The starter system was then checked and the dome fitted. As the intake check was being carried out another dome was found in the intake up in the No. 2 area; we had forgotten to remove it when the old engine was taken out. To get the dome out the radar bullet had to be removed, which cost the team quite a few beers for the 'Fairies' (Radar engineers), and it took a long time for our team to live it down.

After every sortie the starter system is replenished with Avpin, which is a clear liquid. During yet another exercise eight aircraft landed and needed turning (flight servicing), and we were in gas masks and NBC gear and running around like blue-arsed flies. The Avpin was kept in a container that looked like a watering-can. The aircraft were turned and scrambled but four didn't start—much to the disappointment of the pilots. We investigated and found that the Avpin systems had been replenished with water, which no-one had noticed because we were all wearing gas masks and hadn't noticed the lack of smell.

A few of the other incidents were more serious. XR711 'A' had had its reheat system disabled because of a fault so that the aircraft could still be used but reheat was not supposed to be engaged. On this particular day, 'A' was part of a pairs take-off and both aircraft lined up and accelerated down the runway. As they started to lift off, reheat was selected on 'A': the nozzles went to Open and because there was no reheat thrust was lost, the undercarriage was already retracting and the aircraft sank back on to the runway, landed on the ventral tank and engaged the barrier. The barrier stopped him but stretched well over the piano keys. The pilot got out and we dug out the plane but it was too badly damaged to be fixed.

XR711 'A' of 111 Squadron which crashed on the runway at Wattisham on 29th October 1971. [MAP]

Another incident involved XP752 'D' which was part of a four-ship, two Lightnings and two Mirages, while on detachment to Colmar in France. During the flight one of the Mirages turned the wrong way and hit 'D' on the starboard side with his ventral tank. The Mirage landed safely, but the Lightning lost the No. 1 engine, the instruments and part of his canopy due to debris. Despite this, the Lightning pilot too managed a safe landing although the aircraft was so badly damaged that I believe the aircraft was taken to Coltishall and scrapped.

AIR-TO-AIR GUNNERY

Flight Lieutenant MIKE HALE
56 Squadron Tornado F.3 QFI/CIRE;
Formerly of XI Lightning Squadron

Air-to-air gunnery, the stuff of schoolboy dreams, the high-point of so many Hollywood films, the theme of countless computer games and often referred to as the ultimate eye-hand coordination exercise, is for most fighter pilots the sport of kings. Air-to-air gunnery is a skill that is simulated almost daily during mock combat; however, to prove the overall system, pilot/gun/aircraft and to experience live gun firing, air-to-air gunnery is carried out during regular Armament Practice Camps (APC's). APC is the formal name given to a yearly Squadron detachment dedicated to qualify, or requalify, all the pilots in the black art of air-to-air gunnery.

Tanking to Cyprus with Victors in 1985; photo taken by Mike Hale from XS904 'BM' [Mike Hale]

The setting up of an APC was quite involved: the aircraft had to be harmonised such that guns and sighting systems pointed where they should, Canberras booked to tow the target banner ('flag'), ranges booked, tons of ammunition prepared and painted and of course hours of insomnia-curing lectures on the gun, gunsight and range procedures by the weapons instructors. All this work could be spoilt by a couple of weeks of poor English weather. Of course the gun could be used in even the worst of weather, the Lightning was an all-weather fighter, but for safety and clear range procedures good weather was required. To guarantee good weather, Lightning squadrons would deploy to Luqa in Malta and (from the late seventies) Akrotiri in Cyprus. Well that's the official story—lots of sun, sea, sand, pretty girls, cheap booze and monster kebabs probably had more than a passing influence.

If you mention air-to-air gunnery to a Lightning pilot, he will instinctively lean his head and shoulders slightly to the left, bring his right hand up as if holding a control column, start to lean backwards, draw his raised hand rearward and twitch with his index finger with a dreamy look in his eyes and a broad grin on his mouth. He will be lost to intelligent conversation for a while as he day-dreams of his best gunnery pass, counting holes in the flag, barbecues on the beach and all night Summer Balls at the Akrotiri Officer's Mess. On a more serious note, gunnery really is a black art; last year's "Top Gun" can be this year's worst shot and vice-versa.

Concerning air combat, it was often said in the single-seat fighter world: "the first man to the chalk board will win the fight because the sortie will be debriefed from his view-point". Gunnery was different since every sortie resulted in a black and white score, percentage of total rounds fired versus those on (through) the flag. No arguments, no excuses, your score was your score—end of story. Moreover, that score would be appended to a large scoreboard kept on public display outside the Squadron. There it would be on full view to all the Squadron personnel, Canberra crews, officers of the host station, official and personal visitors—very difficult to convince that pretty little nurse you're a star if you've got lousy scores posted!

The Lightning guns and sighting system were designed and set for high explosive (HE) ammunition to be fired. For practice firings inert 'ball' ammunition was used since ball rounds exhibit similar ballistic characteristics to those of HE. (Despite their name, ball rounds are not ball-bearing shaped!) At the time in question, 1984, the RAF started to run out of ball rounds and the remaining stocks were to be saved for use by the Tactical Weapon Units to train student pilots.

The alternative of using HE rounds for APC, whilst a most amusing thought for the junior pilots, was considered too dangerous. Not only would ground handling become a nightmare—every round was effectively a grenade—but airborne use would be fraught with problems too. What if a shell hit the solid part of the flag and exploded, or an errant round exploded close below or behind the Canberra? (Real kills don't count.) And where would they end up? If they didn't explode as they entered the sea they would end up sitting live on the sea bed—anyone for scuba diving? So the powers-that-be decided armour piercing (AP) rounds, as used by the mud-movers (Jaguars, Harriers, Tornadoes etc.), would be allocated to the Lightnings for APC's. The allocated AP rounds were of the solid slug type with a light aluminium case surrounding a solid inert super-heavy centre. As was to be expected, the AP rounds, being much heavier than ball or HE, exhibited totally different ballistic characteristics. In a nutshell, their muzzle velocity was lower but their speed decayed more slowly—almost compensating for errors but not quite.

Strike Command would not allocate any external resources to calculate or trial the necessary corrections to the Lightning sighting systems. The engineers at Binbrook elected to leave the existing settings in the system, relying on the compensating errors theory. In the light of this, 5 and XI Squadrons decided to run a mini APC at Binbrook to prove the system. A weapon instructor from each squadron was chosen to run the trial, Ian Howe from XI Squadron and Chris Allen from 5 Squadron. Two pilots from each squadron were chosen to fly the actual trial sorties; from XI Squadron, John Spencer (the Boss) and myself; from 5 Squadron, Chris Stevens and Dave Carden. Aircraft from each squadron were harmonised, on XI Squadron XS919 and XS933 were chosen. Just after the trial started, XS904 arrived on XI Squadron, was harmonised and joined the trial. At this stage XS904 had yet to be painted in XI Squadron colours.

As the trial progressed it was noticed that the majority of the hits were well towards the rear of the flag, even with the pilot typically tracking (aiming at) the spreader bar at the front of the flag. Obviously, the deflection given by the sight was not enough for the slower AP rounds. It was tried bringing the firing range bracket down from 400-350yds to 375-325yds and then 350-300yds, thus reducing the required deflection. This did not improve matters much and caused pilots to get close to the scrub/foul line of 250yds whilst still firing (the Lightning was through the firing bracket in about half a second). The engineers were loath to mess about, as they called it, with the sight settings so Ian Howe and myself had a look ourselves. Between Ian's weapons instructor knowledge and my applied physics degree we were able to stumble our way through the calculations and settings and discover that the "gravity drop" correction had been allowed for twice and was only applied at right angles to the plane of the wings irrespective of the attitude/bank angle of the aircraft.

This double gravity-drop error, which apparently had been incorporated in the sight since the Mk6 was introduced, shed light on some Lightning gun-firing myths and old wives tales. A typical one was: "it is better to fire slightly out of range and fire slightly line high". Notwithstanding the efforts of the weapons instructors to put down these rumours, they tended to be borne out by results many times over. Since the Lightning on a typical firing pass is at about 60-70° angle of bank and at 2-3 G then it can be seen that these unofficial corrections would tend to compensate for the gravity drop errors. The removal of one of the gravity drops was straight forward; the correct orientation of the remaining gravity drop could not be achieved. The Mk6 aircraft's Light Fighter Sight was a basic visual missile sight, inherited from the Mk3, rigged to work as a simple gun sight and, I understand, inferior to the Pilot Attack Sight fitted to the Mk1and Mk2 Lightnings.

However, the result of simply removing one of the gravity drops, setting a closer firing bracket of 375-325yds, using AP rounds and advising pilots to track the spreader bar was a dramatic increase in the number of hits achieved on the flag. Previously, typical APC scores would be between 5-15% with a good score being around 30%. After this trial, scores of 20-30% became the norm with scores over 60% a regular occurrence. XS904 fared very well during this trial. Even before the sights were corrected she proved to be a consistent firing platform achieving the best scores. XS933 also proved to be an excellent work-horse turning in good results. However, after the corrections were applied, XS904 in particular became a superb firing platform and, in the hands of the now well-practised XI Squadron firing team, consistently turned in the best scores. Indeed regular scores of 60-90% were achieved, setting a standard that I believe was not matched before the Lightning left RAF service.

John Spencer taxies out in XR728 'BA' to lead the Diamond Nine formation at Akrotiri in 1982.

[Mike Hale]

(Right) Mike Hale in XS904 'BM' lines up as No. 3 at Akrotiri Open Day in 1984

(Below) John Spencer leads the XI Squadron formation at Akrotiri in 1982

[both Mike Hale]

A few months later on, XI Squadron were off to Cyprus. Expectations were running high, we were the first squadron to take the AP rounds on APC and we had two proven good shooting aircraft to boot. In 1984 there were not many tanker aircraft around; Victors and Vulcans were being phased out, VC10s just starting to appear and Tristar tankers a thing of the future. With tanker support scarce it was decided that we would deploy to Cyprus by staging through Europe. Due to the short range of the Lightning, five hops were needed staging through Dijon in France, Pisa in Northern Italy, Brindisi in Southern Italy, Souda Bay in Crete and finally on to Akrotiri in Cyprus. Double night-stops would be taken at Pisa and Souda Bay to allow the engineers, in a following Hercules, to service and maintain the Lightnings. All in all a most excellent way to start a detachment. (Phantoms, and now Tornadoes, fit ferry tanks and do the entire trip in one day with a single refuelling stop.)

The Squadron took ten Lightnings to Cyprus that year, transiting in two flights of five. I flew out in XS904, which was by now my aircraft with the tail letters BM (Big Mother). At that time I was most fortunate to have both XS904 and XR749 with my name on. XR749 was my previous Lightning, a Mk3, soon to be withdrawn from squadron service. The A Flight formation consisted of Mal Godfrey (Flt Cdr), Big John Aldington, Steve Bridger, Craig Penrice and myself. We each led a stage of the transit and I led from Pisa to Brindisi. Unfortunately, Big Mother's Tacan and radar had been playing up (the only real navigation aids carried) so I swapped aircraft with Steve Bridger, taking his aircraft to lead the Flight for this stage.

The Lightning was always short of fuel and staging through Europe was no exception especially when using civilian airfields. We simply did not carry the expected 20 minute holding, plus full instrument approach, plus diversion fuel. Charging into a strange airfield, talking to a controller you didn't understand, trying to fly accurately and smoothly for the chaps formating on your wing and keeping an ever watchful eye on your fuel was challenging to say the least. A call from your foreign controller of "Who cleared you from 10,000 ft into the landing pattern" would often be answered "My fuel gauges mate". Needless to say, during the lean tanker years of 1982-85 Lightnings frequently staged to Cyprus and back without accident or diplomatic incident (which is more than can be said for the longer-ranged and better navigationally equipped Phantom).

Once at Akrotiri the aircraft would be handed over to the engineers for a few days whilst the pilots did the rounds of the "Benos Briefs". So called since the arrival briefings given by Akrotiri's Station Commander, OC Operations, Sqn Ldr Ops, Senior ATC Officer, Security Officer and Doctor would consist of "there will be-no this, be-no that and certainly be-no the other".

5 Squadron's last APC in July 1987, preparing for the final formation flypast led by XS935 'AK'. Note the lack of missiles and the smoke-blackened gun ports. (L-R XS935, XS922, XS901, XS928, XS932, XR727, XS936 and XS458)
[Steve leCount]

This year things were different since Akrotiri Open Day was a few days hence and XI were to provide a Diamond Nine flypast. The couple of days acclimatisation and aircraft servicing were replaced with formation practice sorties from Day One. The Boss decided that far from doing a simple flypast he would wheel the formation around a bit and coordinate this with some high speed passes by a singleton Lightning (John Spencer believed in doing a job properly, leading from the front and that XI Squadron was by far the best Squadron in the RAF—an infectious attitude that rubbed off on all the Squadron pilots). For this we would need all ten aircraft serviceable and three sorties practice. During these practice sorties and the display I flew XS904 in the number seven position. Although the Diamond Nine is a fairly standard formation the position numbering can vary—as you can guess this has potential for confusion.

During the first practice, after the traditional multi-Lightning take-off, rush around and noisy vertical climb, we had formed up into our Diamond Nine over the Mediterranean. Chris Lawrence was flying above the formation acting as "whipper in", making sure the formation was symmetrical and looking good so everyone could pick his position references. After getting the front rows sorted out he called "seven move forward" so I closed on the aircraft in front. He then called "seven move forward and eight move back" so I moved forward more and Chris Jenkinson beside me moved back. This call was made several times until I was almost inside the jet pipes of the aircraft in front and Chris was rapidly backing out of the formation. There then followed: "Ah—sorry chaps, as you were, seven ease back a bit and eight move up a lot and we'll start again". Once the Diamond Nine positioning was sorted out the Boss "worked the formation up" at height using 3-4 G during flat turns, wing-overs and reversals. Then, when he was happy with our performance, he started rehearsing over Akrotiri itself. Down over the field the Boss only used 2-3 G stating that he worked us to higher G when practising at height so that his Squadron would be perfect on the day. In so far as XI Squadron was not the Red Arrows, the Lightning was not a Hawk and we were fighter pilots not display pilots, it was hard but enjoyable and satisfying work. With three days practice under our belt the actual display went very well; the XI Squadron APC was off to a good start with plenty of enthusiastic support from Akrotiri.

The Squadron now settled down to the serious job of qualifying its pilots in gunnery. To qualify, a pilot was required to achieve at least two scores of 15% or more during six consecutive sorties. Sounds quite easy but was far from it. There were three types of sortie, Academic, Lim Op and Op. Initially academic 'cine' sorties were flown where only the gun-camera was used. A weapons instructor would assess the films until he was happy with the individual pilot's pattern flying, flag-tracking, angle-off and range estimation, the pilot would then be cleared "live". Although the ability to track the flag smoothly with the gun-sight was number one priority, the ability to achieve acceptable open-fire ranges and angle-off positioning could make the difference between a complete miss and peppering the flag with holes. Estimating ranges against another fighter is normally fairly straight forward; a typical fighter is about 60ft long with a 40ft wingspan and effectively fills the gun-sight at firing ranges. The flag, on the other hand, is only 6ft by 24ft and presents a 6ft by 8ft target from a typical firing position, appearing a little larger than twice the size of the central dot (pipper) in the gun sight. By the time it was three times the size you were at minimum range and at four times you were at foul range. Bearing in mind that the sight only predicted properly for the calibrated ranges, and to go from open-fire range to foul range took a little over a second, accurate range estimation can be seen as crucial for successful gunnery. Arriving at an acceptable angle-off was achieved by the Canberra turning and the Lightning using his AI.23 radar to start his pass from an accurate range and position behind the Canberra. Too little angle-off (line-astern the flag) meant no flag to shoot at and too dangerous for the Canberra. Too much angle-off meant a larger target but a high G turn to follow the flag, thus making tracking very hard. A comfortable and safe angle-off was about 20°, a good pattern and an accurate tip-in would achieve this.

The academic pattern involved two Lightnings working as a pair, and a Canberra flying an elongated figure-of-eight turning alternately left and right. The Lightnings would then fire during the turns making alternate left and right firing passes with one Lightning always leading the left pass and the other leading the right. Normally two pilots would fly as a constituted pair during this phase, enabling them to match their pattern-flying and thus maximising their allocated range slot.

ACADEMIC FIRING PATTERN

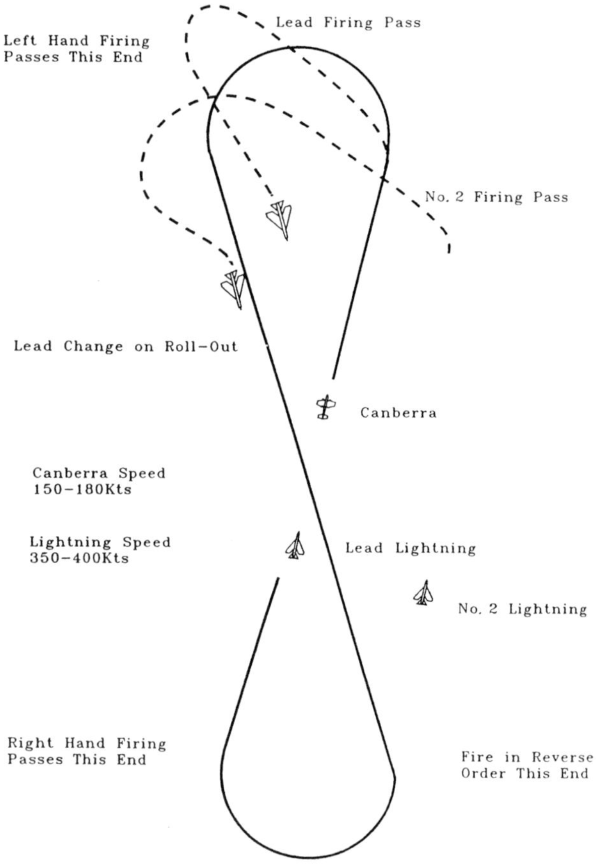

For an academic shoot the aircraft would be loaded with 120 rounds (60 per gun) enough for about six half-second firing bursts. During a range slot ten to twenty passes would be made on the flag. The pilot could take a couple of dry runs to get his eye in, then be selective firing only on those passes he considered to be good. Once qualified, a pilot moved on to the Lim-Op shoot of which only one per pilot was flown. In this shoot the aircraft would be loaded with 40 rounds (20 per gun), enough for two to three quick bursts, and the pilot would only be allowed 4 passes. This forced the pilot to make all passes good, salvaging those passes which he otherwise might not consider firing on. Most pilots preferred to fire in left pass (in the same way they prefer left-hand circuits, formating from their left, high performance turning left, etc.). So for the Lim-Op shoot most pilots had the Canberra flying a left hand race-track producing left-hand firing passes. The Lim-Op shoot was in preparation for the Op shoot, more of which later.

A typical range sortie would start with a brief by the weapons instructor. He would have looked at your previous sortie's gun-sight films, assessed and analysed your performance and be giving you advice on how to improve your scores. At the Ops desk you would check your range slot, Canberra call-sign, allocated Lightning, colour of rounds loaded as well as the more usual sortie domestic details, i.e. diversions, fuel states, etc. Working in pairs, two, four or six aircraft would work one Canberra range slot with each Lightning firing rounds dipped in a different coloured paint. You walk out to the line hut with your number two talking about anything except gunnery. Once in the line hut you check in the 700 (aircraft log book and daily servicing record) and recheck number of rounds loaded and colour. At the aircraft the guns are loaded but the gun doors are open and the gun leads unplugged. Once again you check colour and number of rounds and inspect your aircraft. Before you strap-in or fire-up the beast, you have the guns plugged and gun doors closed. Whilst this is being done you must sit with your hands visible outside the cockpit well away from all weapons switches (unless you want a one-sided conversation with an irate armourer). If you are first on the range you wait for the Canberra to get airborne. The APC dispersal is about 100yds away from the end of the runway and it is not worth wasting fuel if he is late. As soon as the Canberra staggers airborne with his flag ATC will call, and you can see if the flag is flying properly. If it is, scramble start, taxi, airborne and you are with the Canberra as he enters the range. Normally the Canberra will fly at 5000 ft. Low enough for the pilot to see small boats as well as the larger ships and high enough to clear well ahead. Olympus Radar (on top of the mountain of the same name) and/or Garter Radar (Akrotiri radar) will also be clearing the range. You quickly settle into the pattern with your number two since the pair of you will only be allowed 20 minutes total time "on the flag".

You approach the Canberra from above and behind at approximately double his speed. Your number two will be swept out about a mile in your four or eight o'clock. At a mile and a half behind the flag you call the Canberra to "commence" and he enters a turn. Gunsight on, radar transmitter off, fold the radar boot away and look for the flag as the Canberra starts to build up angle-off. Judge your moment, tip-in, call "in-hot" and listen for clearance to fire, trigger safety catch off, start bringing the sight up behind the flag, quick glance down at the speed and adjust power as necessary, is the angle-off building up nicely? No—terminate pass, yes—continue, bring the pipper on to the flag as the flag starts to grow, G building up, anti-G trousers starting to inflate, must be steady now—sight needs time to settle, pipper same size as flag (1000 yds), too late for anything but the smallest corrections, flag twice the size of the pipper (500 yds), half-second pause, hold breath, squeeze trigger, guns fire, release trigger, immediately ease back on stick plus a boot-full of top rudder, barrel over the top of the flag (don't go below you might collect debris). Out of the corner of your eye you see the flag ripple and the spreader bar jerk, YES!! hits on, trigger safe, call "out" so your number two can be cleared in, pitch up into wing-over, radar on, look back over your shoulder to see your number two completing his pass. As he pitches up you now follow him into the next pass.

Both of you now glue your eyes on to the Canberra, who is now below and behind you heading in the opposite direction and disappearing in the Mediterranean haze. Ignoring your instruments and flying by feel, you drop down behind the Canberra until you are once again line astern and closing rapidly. Keeping your eyes on the Canberra caters for the occasions when the radar fails and saves you the embarrassment of losing it. Sometimes the summer haze would be so dense that you had difficulty seeing the Canberra outside of a few miles and could not pick out the flag until well into the firing pass. On these occasions you would need the full range slot to achieve enough good passes. However it was more usual for you both to have fired-out and have a few minutes in which to inspect the flag and do a quick count up before the next pair arrived. Red and blue were always the best colours, with the rounds making brightly coloured holes, whereas green, purple and yellow holes would be more difficult to see. This year I had been fortunate and had done most of my firing in my own aircraft, XS904, achieving some good scores despite shooting the flag off in my first live sortie. A round went right through the metal coupling that joins the towing-strop to the spreader bar. It certainly improves your break-out technique when the flag stops dead in the air right in front of you. During the early stages of the APC the Squadron would be working hard and fast, generating armed aircraft to meet continuous range slots throughout the day; so, once finished on the range, and to keep the engineers happy, it would be a high-speed dash back to Akrotiri, being careful not to drop a sonic boom on the way.

Checking the banner at a mini-APC , Binbrook 1985:
(L-R) John Aldington, Chris Jenkinson, Martin Ramsay, Craig Penrice, Bob Bees, John May, Chris Lawrence (kneeling), Ian Hollingworth and Simon Bostock. [Iain Small]

At the end of a range slot the Canberra would drop the flag back at Akrotiri and its collection and delivery to the Squadron would be eagerly awaited. The duty weapons instructor by now had usually debriefed the films of those firing on the flag and, if he felt confident, would predict the scores of the various pilots. If six aircraft had been firing on the flag and they had done well the flag would be a myriad of colours. On the other hand, if the flag had only been up for a couple of aircraft who had done poorly then it was not unknown for the flag to come back "Virgin". The weapons instructor would then score the flag (allocate hits for simple holes, multiple/jagged holes, coloured streaks, marks and general damage to the flag and spreader bar). A hundred or so 30mm rounds can make a mess of the flag and in this, the most emotive part of the APC, there was obviously room for debate and disagreement. Different instructors had their own styles. Steve Bridger, for instance, would shuffle up and down the flag, announce a score and depart, no discussion no appeals. Conversely a flag assessed by Ian Howe would be accompanied by much ranting, raving, screaming "you cannot be serious", threatening, waving of arms and sulking before a score would be declared. Some colours such as yellow and brown or blue and purple leave similar marks, especially if it is just a nick at the edge of the flag or in the dark frayed tail end. Sometimes two pilots (say flight commander and junior pilot) might just need one more hole to qualify and they are both looking at the same mark each claiming the hit and the instructor has to make a decision. On one side there is the flight commander stating that, not wishing to influence anything, but he would shortly be writing the instructor's annual assessment. On the other side is the junior pilot chorus-line loudly supporting their man and threatening the instructor with a seeing to behind the bike sheds. Not a good day to be duty weapons instructor!

The final stage of gunnery training would be the Op-Shoot. In this phase the Squadron as well as the pilot would be assessed. Once the pilot had signed and walked for his aircraft there was no turning back. If the Lightning went unserviceable before or after take-off, or the pilot could not find the Canberra or generate safe firing passes, or the guns failed to fire or jammed, or in fact anything that could be put down to the Squadron, would not invalidate the sortie which would be scored out of the 40 rounds loaded. The Op-Shoot followed a different format to the academic and Lim-Op patterns, it was basically a race against time. In the Op--Shoot the Lightning met the Canberra head-on in the classic "beak-to-beak" neutral start to air combat guaranteed to get the adrenaline flowing. You would work alone having the Canberra to yourself. Most pilots liked to reduce the fuel load and thus the weight of their aircraft before commencing an Op-Shoot. There was not the facility to dump fuel in the Lightning, apart from the two massive reheats at the back, and that is exactly what was used during full-burner vertical climbs after take-off. Unfortunately, directly under these climbs was the Akrotiri police piquet post who retaliated by stopping and thoroughly searching all XI Squadron vehicles returning from town in the evenings. For myself, I preferred to keep the aircraft heavy, and therefore less twitchy, and just potter along with the Canberra looking for a clear area in the range.

OP-SHOOT PATTERN

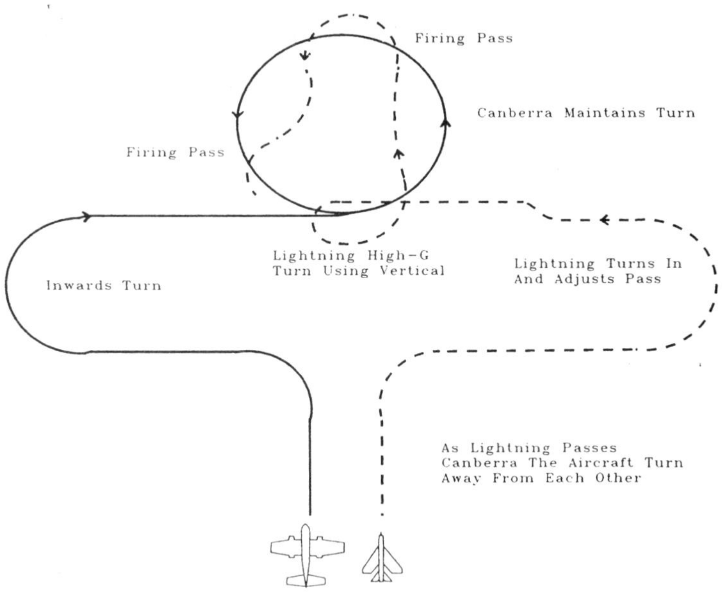

An Op-Shoot starts with you flying alongside the Canberra, and on your call of "outwards turn", you both turn 90° away from one another and head off in opposite directions. After about 30 seconds you are about five miles apart and you call "inwards turn". You now both turn back in through 180° and the Canberra will fly in a straight line; you are now flying straight towards each other. On a typical Mediterranean hazy day you will pick the Canberra up on radar first, then visually at only two-three miles. You start positioning to pass him such that the subsequent direction of turn, and firing passes, will be left or right depending on your preference. If you pass too far from the Canberra he will call "foul" since you have cheated and given yourself too much turning room; if you pass too close you will frighten him and he will call "Stop" since you have infringed separation minima, either way you lose. As you pass the Canberra he will turn towards you and continue turning for 540° (one and a half turns). At the pass you have basically three options: delay for a few seconds then a maximum performance reheated turn to drive into the Canberra's six o'clock and start your pass from the traditional academic position; a hard full cold turn into a lag pursuit position outside the Canberra's turning circle then pull inside the circle to commence your pass; a pitch up into a wing-over behind the Canberra ('on to the perch') before spiralling down inside his turn to commence your pass. With enough rounds for two-three bursts (or one big one!) you then follow the Canberra around his turn diving in and out of his circle generating safe firing passes. Once the Canberra completes his 540° of turn he will call 'Stop' and roll out. That would be the end of the Op-Shoot, only one per sortie and only two sorties per pilot.

As you turned outwards away from the Canberra the game was on. It was too late to call stop to reposition, restart, sort yourself or your aircraft out. The only way the Op-Shoot could be restarted was for the Canberra or Olympus Radar to call "range foul", i.e. another aircraft or perhaps a boat getting too close. You set the radar up (range, elevation, gain etc.) and, as you turn-in, the Canberra marches onto the side of your scope at about six miles. You centre him up (point straight at him) and start peering through the haze for the Canberra. Searching, searching, getting close now, shadow in the haze—there he is, quick jink to sweeten up the pass, radar off, gunsight on, radar boot away, take a deep breath, stay calm, merge with the Canberra, stick hard back, that lovely swept wing roar as the wings bite the air, 40° nose up, drop left wing to keep the Canberra visual, overbank and pull around above and across the Canberra's tail, tap the blowers (minimum reheat) to stop the speed getting too low, play bank-pull-power

(Top) XS932 'AG' of 5 Squadron taxies out for a sortie at Akrotiri in July 1987. [Steve leCount]
(Middle) Mike Hale beats up the pan in XS904 'BM' at Akrotiri in 1984. [via Mike Hale]
(Bottom) During APC in 1985 three aircraft of 5 Squadron were decorated with 'sharkmouth' markings; shown here is XR754 'AE' at Binbrook on 2nd April 1985. The other aircraft were XR770 & XS903. [HT]

to arrive on the perch, cancel blowers, set cruise power; you've cracked it—only got to hit the flag now! The Canberra has turned through 180° and you have about 360° of his turn left to get in two firing passes. Double check guns selected, sight on and radar off, pick your moment, tip-in, check speed and power, look for flag, call "in hot", track flag being more positive with your corrections, you must make the pass a good one, is the pass safe? is it a good pass? then give it a long burst, average pass? normal burst, hopeless pass? don't fire, make trigger safe to resist temptation to waste rounds. As you break-out assess the situation. How far has the Canberra to go before calling "stop"? How many rounds left? - they will all count against you if you take them back. Deep breath, shake the sweat out of your eyes, wing-over straight back in, trigger safety off, think options—lots of rounds left? then open slightly out of range for long burst, last pass not so good? this one must be the best. In, fire, out, all weapon switches safe, game over, relax.

Unlike the academic phase, after an Op-Shoot there was not normally the rush to get the aircraft back for the engineers. You now had a Lightning with both time and fuel to yourself. No number two to think about, no leader to follow, no nagging navigators, and the most exhilarating aircraft just to enjoy for the sheer pleasure of flying. There were many options: a visit to the "Cyprus Buoy", a well known anchorage for Soviet Bloc warships; a maximum reheat acceleration at low level (300 to 700 mph in a few seconds) followed by a vertical climb (40,000 feet in less than a minute) watching the altimeter become a blur; a Mach 2+ run, followed by a ballistic climb; or indulge in a little "home movies"—use the cine button to put something on the end of the gunsight film to entertain the weapons instructor. The list was endless.

Due to good fortune, and after much nagging, pleading, threatening and bribing, I had managed to do the majority of my firing in my own aircraft, XS904. When all the shooting, scoring, arguing, sulking, counting and averaging was complete I had just managed to beat Ian Howe to gain the Top Gun prize. In contrast to previous years, where a considerable amount of time and effort was put into qualifying pilots, this year the Squadron qualified its pilots quickly and we had plenty of time left before returning to the UK. Consequently, the Boss decided that we would put up another Diamond Nine formation and perform flypasts at Akrotiri, TPMH (military hospital) and Episkopi (Headquarters Cyprus). Once again I flew in XS904, but this time in the number five position (outside left). Here it was easier than in the number seven position I flew in earlier (seven, eight and nine were often referred to as "flak in action" down the back end). The format for the flypasts included the traditional stream takeoff and "follow me" rush around the airfield at low level before departing in a vertical climb. After the flypasts the recovery broke with tradition. The multi-echelon run-in-and-break was followed by yet another "follow me" low level rush around the airfield before reforming in Diamond Nine. The final run-in-and-break was then flown in Diamond Nine with the aircraft peeling off one-by-one to land. A nice way to end the APC before handing the aircraft over to the engineers to prepare for the trip home (and indulge in a few mega kebabs and bottles of kokinelli).

When it was time to come home the Squadron was offered tanker support in the shape of Victors to overhead Sigonella (Sicily) and then the new VC10s to the UK. Most of us at that stage had not tanked from the VC10 but their support was accepted. The Victor, Vulcan and Hercules tankers we were used to had solid wings that stayed horizontal when they flew, providing an excellent formation reference for tanking. Whereas, the VC10 wings bent up and flapped around when the aircraft was flying, particularly when in the turbulent air over the Alps. You could see the "thinks bubbles" coming out of the cockpits as the chaps moved forward in turn to tank. However, after a five hour transit, we arrived back at Binbrook and a most enjoyable APC came to an end. I look back on those sorties in XS904, firing on the flag in the Cyprus sun, as possibly the most enjoyable flying I am ever likely to experience. Now, as an instructor on 56 Squadron Tornadoes, and in this age of all-aspect radar and heat-seeking missiles, I have just been credited with a "guns kill" on my current Boss in the middle of the ACMI combat range—so I must have learnt something in XS904 ten years ago!

ARMOURERS ON APC

Chief Technician MICK CAMERON AM Inst. BE
Formerly of XI Lightning Squadron

Armament Practice Camps or APC's involved a fair amount of preparation especially for the armourers. New, serviced guns had to be fitted and harmonised, normally this went on long into the night. The overwing tanks had to be fitted and the riggers usually had a hand in this. When fitted, the tanks had then to be torqued down. This involved two armourers using a six-foot torque wrench with a six-foot extension bar attached. If I remember correctly, the torque setting was around 460 lbs ft! This certainly built up arm and back muscles, but the 'exciting' bit was when the torque wrench 'broke'. Each person used to gingerly back off the torque wrench slowly, hopefully controlling the point when the monster reasserted itself. I can't think of many plumbers who haven't received a faceful of 'Britool'. Luckily the suntan soon covered the bruises!

For Malta APC's, the missiles were sometimes removed in the UK as there was limited storage for them on the island, then it was a case of packing all the tool kits with essentials such as flippers, snorkels and flip-flops, loading the aircraft and away.

The choice of transport made a difference as to how the detachment started, VC.10's were fast and cramped, Britannias were slow and cramped, Herc's were slow, cramped, noisy and cold, but a Belfast was very nice thank you. Not too fast, but on the upper passenger deck it was very comfortable.

After the arrival, moving in, unpacking and then recovering from the first hangover, it was down to business. The overwings had to come off first, and then the familiarisation and camera sorties were carried out. The bomb dump had prepared the rounds for firing by dipping the bullet heads in non-curing paints so as to leave a coloured mark as they went through (with any luck!) the banner.[1]

The aircraft for firing had various pre-firing checks carried out and grease smeared around the blast deflectors and gun muzzles. This would help prevent the carbon from sticking to the aircraft. The ballast-filled ammunition tank was then removed ready for uploading live ammunition.

This was achieved by taking tension on the tank hoists and then removing the tank hatch from under the ventral. One never used to go underneath the hatch when undoing the securing latches - just in case. 240 rounds of ballast would certainly cure any hangover! The ballast tank was then wound down and removed. Simple 'loading' straps were then passed from inside the ammo tank cavity through the feed chutes and feed mechanisms to dangle out of the gun pack doors. A 'live' ammo tank was positioned under the ventral, and the belts of ammunition were connected to the ends of the straps still left hanging in the tank cavity. One man would then lie under the aircraft on his back (in all the fuel and other sticky stuff!), take hold of the free loading strap ends and keep tension on them. Another two persons would wind up on the ammo tank hoists, and the man on the ground would 'jiggle' the tank into the ammo tank cavity with his feet while keeping tension on the straps to prevent the rounds falling back in the tank. With the tank fully up and the hatch back on, the loading straps were given a good tug and hopefully the rounds would pull through the feed chutes and just start to enter the feed mechanism. If this didn't all go according to plan, a prod with a screwdriver usually cured the problem. If the rounds still wouldn't budge, then lowering the tank a tadge usually sorted the problem out. If the aircraft was going to fire straight away, a four bottle gas trolley with an Aden gun cocking hose adapter was used to cock three rounds into

---

[1] *Of course the groundcrew used to 'amend' the banners dropped by the Canberra after a gun-firing sortie. Sometimes a well and truly plastered banner would be replaced by a virginal new one. The look on the face of the budding Tom Cruise was a picture when he tried to count his rounds that he was sure had hit. Another common ploy was for the armourers to lay the banner on the ground and push rounds through to simulate lots of hits, and then show it to the pilot who thought he had just completed a camera sortie! Another trick was to replace a banner with one that had red and yellow hit markings and then show it to two pilots who thought they had used blue and brown tipped rounds respectively. Finally, I remember one enterprising pilot who only fired one round before a gun jam. The round hit the banner and he tried to claim 100% hits......*

the gun, the first round going into the firing chamber. All we needed then was a person in green to borrow our jet for a bit.

After start-up the gun electrical leads were connected and the gun bay doors secured. The aircraft then taxied to the end of the runway, where an extremely tanned armourer would remove the armament safety break, making the weapon circuits live. After they were in the air, it was time for a well-earned 'Coke' and make a few plans for that night's entertainment.

On landing, the aircraft taxied up to the now even darker armourer at the end of the runway for the safety break to be plugged in,

Luqa 1970: XS932 'H' and XR763 'B' land and taxi in [Both Crown Copyright/MOD]

and then taxi back to the pan. A good indication whether the guns functioned as advertised would be the amount of brass visible through ventilation holes in the rear of the gun bay doors. If all went well, another tank of ammunition was loaded ready for the next sortie. If things were not so good, it could be anything from a quick fix to a gun change, depending on the damage.

This blissful existence could go on for up to six weeks, and then unfortunately it was back to blustery Binbrook, probably by a Britannia diverted to Brize Norton with the final leg by RAF bus![2]

---

2 *Seatstar is a company run by Mick Cameron, Vince Draper and Gary Milton which undertakes servicing, technical assistance and refurbishment of ejection seats. Mick and Gary made safe the ejection seat and other explosive cartridges on XS904 upon her delivery to the LPG. Tel.: Mick 01638-508294; Gary 01480-412630; Vince 01553-671438.*

SAUDI LIGHTNINGS RETURN

Group Captain D. A. WILLIAMS AFC RAF
Detachment Commander, Italian Tornado Leasing Team, Gioia del Colle;
Formerly of XI, 23 & 29 Lightning Squadrons,
and CO of 5 Lightning Squadron

One late autumn afternoon in 1985, I was reflecting on my first few weeks in command of No.5(F) Squadron. I remember thinking how lucky I was to return to an aircraft that I had enjoyed flying so much in the past. At the time, I had accumulated 2500 hours on seven Marks of Lightning and the thought of reaching the magic 3000 on type really excited me. At the end of my tour, the squadron would re-equip with the Tornado ADV, so this was the last chance to 'fly my pants off'.

My all-too-rare moment of peace was disturbed by my Adjutant: "The Station Commander wants to see you urgently." What had I done, or not done! I couldn't think of anything. Intrigued, I made my way to his office in SHQ. His PA greeted me with her customary smile. "What's this all about?" I asked inquisitively. "Can't say!" she replied. By now I began to feel a little uneasy, because the PA usually dropped the odd hint. This time her lips were sealed. "The Station Commander will see you now".

I entered his office, and was invited to sit down. "Have a cup of tea," he asked in a cheery voice. The relief of that invitation was music to my ears, for I knew the man's style. Had my visit been for a one-sided interview, I certainly would not have been given the opportunity to sit down, let alone be offered a cuppa! "Andy, I've just heard from HQ 11 Group that the RAF has been tasked to recover 22 Lightnings to Warton from Saudi Arabia. I've asked O.C. Operations Wing (OCOW) to lead on a number of administrative matters, such as how to get you to the Middle East, accommodation, visas, etc., but I want you to look after the flying side."

"When is this likely to happen?" I asked. "We're not sure exactly, but STC seems to think early January." "That's six weeks away, no problem sir." I returned to the squadron.

Many years previously, I had been a weapons instructor on the Lightning Conversion Unit (LCU) at Dhahran. During my three-year tour I had also flown from Riyadh and Khamis Mushayt. The latter airfield, at a height of 6800 feet asl, offered many 'gotchas' for the unwary pilot. I recall in the middle of the downwind leg on RW27 that one had to weave around vultures soaring above a waste tip behind the Junior Mess. Thankfully, I never had a bird-strike; had this happened, I believe the mighty Avon 302 engine would have coped, unlike some of the rather delicate engines in modern fighters. I never had the opportunity to visit Tabuk in the northwest. In the mid-to-late 70s, the Lightnings had deployed there from Dhahran to make space for F-15s, following an RSAF re-equipment programme. I was looking forward to returning to Saudi Arabia as a number of colleagues were employed by BAC; it would be nice to see them again. My real interest though was knowing that the Lightnings would be returning to the UK and not left to rot in the desert like the Kuwait Lightnings. But why were the aircraft required back in the UK? Could it be that the MOD had decided the aircraft was about to get another lease of life? After all, we had been round this buoy before. There had been plans to phase the Lightning out of service in the mid 70s; indeed the OCU at RAF Coltishall closed in 1974, but almost immediately re-opened as the Lightning Training Flight at RAF Binbrook.

Soon all would become clear: this was not to lead to another in-service extension of a 'very tired old lady', but was part of a commercial deal being hatched between British Aerospace (BAe) and a European country in need of a short-range high-performance interceptor. However, that's another story! The upshot was that BAe wanted 22 Lightnings from the RSAF to refurbish and pass on to their customer. RAF assistance had been requested for the recovery phase, which would involve air-to-air refuelling (AAR). HQSTC had decided to use a mix of VC.10K2/3 and Victor aircraft. The VC.10K had the advantage that it could fly non-stop to Saudi Arabia and carry the Lightning pilots as passengers. My thoughts of travelling first class in a British Airways 747 were short lived!

I wasted no time; with Christmas around the corner I knew the Station would close for two weeks. I sent my AAR officer to Brize Norton to co-ordinate the trail, which would include flying over parts of the Middle East new to us all. My "Gofer" spoke to the Accountant Officer to see if we would be entitled to any allowances.

Nothing substantial, but what more could I have expected! The squadron's combat survival and rescue officer made sure his knowledge of desert survival was up to speed and prepared briefings for the rest of us. I had to find someone to take by hand to the Saudi Arabian Embassy in London passports and other necessary documentation. Although I had lived in Saudi Arabia for three years, I had forgotten how much family history and other personal details one had to own up to before a visa was granted. The navigation officer ordered the maps. We would be flying a distance of some 3000 miles, and therefore required lots of maps! The Lightning cockpit had not been designed to accommodate so much paperwork. Besides, we pilots rarely carried maps because we never flew far enough from Binbrook to need one! However, this trip was to be different and we took our navigational responsibilities very seriously.

Slowly the plan came together. We would recover the twenty-two aircraft in two waves, twelve in the first wave and ten in the second a few days later. We needed to ferry out extra pilots to allow for sickness. We would be bringing home four T55s, the two-seat trainer, and we decided in the planning stage that, whilst this aircraft can be flown safely by one pilot, two would be company! OCOW decided to fly out a few days in advance of the main party to ensure that accommodation and other administrative functions were in place. Meanwhile back at Binbrook, we were holding briefings followed by more briefings. Eventually, in early January 1986 the time had come for the pilots to fly out to Saudi Arabia via Brize Norton.

We were originally scheduled to travel to Brize in an RAF bus, but somehow we managed to persuade No 101 Squadron (the Tankers) to collect us from Binbrook. I was quite surprised that HQ 1 Group had authorised this taxi service; maybe they never knew! We were all pleased that we did not have to suffer a four-hour journey by road! After we had checked into the Gateway Hotel, we went to the Bull in Burford for a few jars of fine English ale before heading off to a curry house. Why on earth we went Indian that night, knowing that within eight hours we would be at FL350 in a VC.10, God alone knows! I enjoyed my vindaloo better the first time around.

Because of the time difference (Tabuk is three hours ahead of GMT) and the high temperatures even in January, we had an early start on the day of our departure. We checked in with No 101 Squadron operations at 0430hrs for take-off one hour later. As the wheels retracted, I nodded off. I woke up some four hours later as we were flying down the centre of the Mediterranean. I must say that the azure blue water looked very inviting, but all I wanted, was to get rid of a Gorilla in my mouth. The loadmaster offered me a coffee, which helped. Just when I was about to nod off again, the captain asked me if I would like to fly the aircraft. I could hardly refuse. As I climbed into his seat, taking care not to stand on the throttles, I immediately felt better. He reminded me that the passengers would not be too pleased if I threw the aircraft around the skies like a fighter. However, I could not resist putting on 30 degrees of right bank and then reversing to 30 degrees the other way; the shouts of derision from the back must have been heard miles away. I was impressed with the rate of roll—I wondered what else I could do! Unfortunately not a lot, as we were supposed to be flying accurately down the airway, so I handed control to the co-pilot and returned to my seat.

We had no clearance to overfly Israel and therefore entered Saudi airspace from Egypt, just south of the Sinai desert. Tabuk came into view. The airfield was much as I had imagined: a large military installation built in the desert; quite a feat of civil engineering. We landed. The Saudi air was no kinder than I remembered from all those years ago—dry, hot and dusty. What I would have given for an ice-cold beer! OCOW met us and clearly had won over the hearts and minds of the immigration and customs officials, because a good deal of the bureaucratic paraphernalia I had experienced in the past was not required. We still had to endure the duty queue in the hot sun, but this time it seemed quite painless. Perhaps the alcohol from the previous night was still anaesthetising my brain!

Eventually, we made our way to the Senior Mess, where we were to spend a couple of days before flying back. I had a light meal (about all I could eat) and, as expected, met up with a number of "ex-pats" whom I had not seen for ages. Later, a small group of us drove into Tabuk town. I was under orders to buy presents for my wife and kids. During my tour on the LCU, I had bought gold, which turned out to be a worthwhile investment. The streets of Tabuk were lined with many jewellery shops all willing to bargain for that 'special rock-bottom price'. I found exactly what I was after, bangles for my daughter, a neck chain for my wife and a couple of audio tapes for my son.

Nine F.53's (53-666 nearest) detached from Dhahran to Riyadh in May 1970 for the King's birthday flypast.
[Falcon Aviation]

Saturday 13 January 1986 arrived and we were up very early in the morning—it hardly seemed as if I had been to bed. OCOW had been at work for a while when we arrived for our briefing, as he had to phone the operations room at Strike Command to receive the Go/Nogo message. This last-minute check was always required on AAR Trails to confirm diplomatic clearances had been received, and that the en-route weather was satisfactory. The news was good - we had a Go. The weather would be fine until Italy and then we could expect a slow deterioration as we headed towards France. The forecast for Warton at the time of our arrival was showers; not what I wanted to hear, but there were plenty of diversions. I gave the formation brief and ran over a number of contingency options. Surprisingly, I had been told not to use Binbrook as a diversion unless in dire emergency. This came as a surprise as our route was to take us within 50 miles of the station—I never have been given an acceptable explanation for this rather odd decision. Our departure from Tabuk had been very carefully thought through because of the proximity of Israel. Behind the scenes, British diplomats had briefed government officials in Tel Aviv of our route, and timings; I suspect the Israeli military were faxed a copy of the flight plan. A large number of unannounced fast-jet aircraft, originating from Saudi Arabia, appearing on Israeli air defence radars may well have sparked off another full-scale war in the Middle East! The three formations of four aircraft (plus AAR support aircraft) were to depart at 45 minute intervals; this spacing would cater for unserviceabilities. I was to lead the first wave. Because we were to arrive in UK in mid-winter, we had to wear immersion suits which, in the heat of the Saudi sun, were very fatiguing. Just before 0900hrs we walked to the aircraft. Knowing that the flight time was to be around 7 hours duration, I had only a small glass of orange juice with my cornflakes—no coffee, as that always went straight through me and would surely create some discomfort shortly after take-off. I had never used a 'pee-bag' in a cockpit before and was not going to start on this trip!

My aircraft looked in pristine condition and glinted in the morning sun. The Saudi Air Force had never toned down their Lightnings. I climbed aboard and to my horror discovered there was a major problem. During my pre-start checks I noticed that the oxygen contents were very low; the LOX had leaked, not an uncommon fault, but it would happen to me. I therefore had to take one of the spares. In itself, this change of aircraft would not normally have been a problem, but wearing winter flying clothing in that heat and moving quickly between aircraft was not the way to begin the long flight home! Feeling rather hot and flustered, I eventually fired Lightning F53, ZF581, into life. The rest of my formation had been waiting patiently for a few minutes. We checked-in with the VC.10K on the radio, and were soon on our way to the take-off point. The tanker took-off before us as it needed time to climb to FL240. We would all meet at about 60 miles from Tabuk. Meanwhile, I had briefed my formation for a 20-second stream take-off, steep climb and a high-speed (0.92M) pass down the flight-line; pretty standard stuff for Lightning pilots! For the flypast, I set the authorised minimum height and maximum speed and I can assure all readers of this article that no rules were broken!

(Top to Bottom) ZF580 (53-672: now gate guard at Samlesbury), ZF594 (53-696: now preserved by the North East Air Museum) and ZF577 (53-668: currently in a container with Portsmouth Marine Salvage) stream their chutes and taxi in at Warton [British Aerospace via Jimmy Dell]

The RV with the tanker and first refuelling bracket were uneventful, but I remember comments from the tanker crews about the extra 1000lbs fuel we had each taken. I put this down to a hiccup in the fuel plan rather than excessive use of reheat during our departure! Our initial track took us on a southwest heading away from Israel, south of Sinai and west towards Egypt. When we finally turned onto a northwesterly heading, I noticed below me an irrigation development, which was very impressive. A finger of rich green land reached into the desert sand. We coasted out of Egypt at Alexandria.

The journey down the Mediterranean towards Sicily went without a hitch. The tanker navigator ensured that we did not stray off our route. As we were out of range of any land-based TACAN beacons, my contribution to the navigation was zero. Lightning pilots learned very early in their lives that once joined to a tanker so far from land to hang on like a limpet, particularly if there was a risk of high level cloud. We had the AI.23 radar, with which we could lock to the tanker, but 'sod's law' usually prevailed during these occasions. Just when you wanted the radar, it would fail!

The VC.10K, which had brought us from Tabuk handed us over to two Victors, which had deployed to Palermo a couple of days earlier. The RV went like clockwork and it was not long before we had our fuel tanks topped up (yet again!), and were on our way to Sardinia, Corsica and up through France. During the 7-hour flight, we had a total of 8 refuelling brackets. People have often asked me what it is like to sit in a tiny cockpit, trussed up like a Christmas turkey, on long flights such as this. Bloody uncomfortable! Seriously, I found that I always had something to occupy my mind. We would use a quiet frequency when we were refuelling, but I would then ask the tanker captain to transfer us to the air traffic frequency that he was working. In doing so, I was able to plot the progress of the route on my map thus ensuring if the worst came to the worst, I would be best placed to find my own way to a diversion if I got separated from the rest of the formation. Thankfully, that never happened.

Six hours after take-off, we were entering UK airspace, and at this point, were released from the tanker. I thanked the crew for their excellent service, increased speed to 0.9M and took over the navigation. The sortie so far had gone according to plan; things were about to change. We made our way towards Warton in and out of cirrus cloud. The formation maintained visual/radar contact with me. Before being handed over to Warton Radar, the temptation to fly through the Binbrook circuit was overwhelming. After all there were some very young Lightning pilots who had never seen the aircraft in its metallic finish. However, I resisted! The Warton controller welcomed us with the airfield state and latest weather: "All aids serviceable, colour state white." "Request actual", I asked. "Latest weather - 2 eighths at 1500ft, 6 eighths at 2000ft, 8 eighths at 2700ft—visibility 25 kms—recent showers—surface wind 340 at 20kts gusting 28."

That was good enough for me to call my formation into close box formation for my briefed flypast. We began our let-down from FL240 at about 35 miles to the southeast of Warton. For the descent, we were in thick cloud. I knew this would present no problems for my experienced mates—they had been in similar situations many times before. We levelled off at 3000ft, ten miles, in dense turbulent cloud on the extended centre-line of Warton's westerly runway. Having heard the main base was at 2700ft, I asked for further descent. The controller cleared me to 2500ft, but we were still in the cloud! The range was down to four miles. I requested a descent to 1500ft. As we levelled off, the ground was just visible. Whew! Bit tight, I thought. I saw the airfield slightly to my right, made an adjustment and flew past the tower at 500ft msd at 350 knots. I could see that there was quite a large reception party to greet us. Hardly surprising as the Lightnings were 'coming home'!

Before we left Tabuk, I had briefed that following the box flypast at Warton we would turn downwind and change formation to low-level battle for a run-and-break to land. As we completed our manoeuvres and lined up for our final run, to my horror I could see out to the west a big shower approaching the airfield. We would make it if I poured on the coals. I increased speed to 500 knots, but as we broke into the circuit I began to sense that all was going to end in disaster. I called the break, pulled 6G, set the throttles to idle, selected airbrakes out, and was willing the speed to reduce to below 250kts so that I could lower the undercarriage and put the aircraft on the ground. It was so dark outside I could see cars with headlights on. As I turned finals, I switched on my instrument lights! The rain had reached the far end of the runway and I was having difficulty seeing the coastline beyond. I knew that I would be on the ground before the airfield disappeared from view, but what of the rest of the formation? I could see the windsock almost horizontal and pointing across the very wet runway. I did not ask the local controller for the surface wind for I knew it was probably outside limits and then I would be forced to hold off or divert. Surprisingly, I felt quite comfortable as I lined up with the runway. I crossed the threshold, flared (I never could crash-land the Lightning, which was the recognised way of landing the aircraft!), and to my great relief, I felt the brake parachute tug as it deployed. I braked gently taking care not to burst a tyre, and as the speed decayed I could feel the increasing effect of the crosswind on the 'chute, which I jettisoned before I yawed off the side of the runway! I looked over my shoulder to see my No. 2 landing behind me, but Nos. 3 and 4 had to overshoot. Momentarily, I thought that I should tell them to divert. I didn't. Although the pressure was on, they were responsible enough and in the best position to make the correct decision; besides they were big boys. I watched them both turn downwind. During his overshoot, No. 3 had asked for the surface wind. The crosswind component was on the limit, but gusting well outside. Sensibly, as he turned finals for the last time, he asked for the surface wind to be read out; if both failed to land off this approach, I knew they would have to divert. My fingers were crossed. Thankfully, they landed safely although I noticed they had difficulty with the wind before they turned off the runway.

By now I had taxied to a waiting marshal. Only a few of those spectators I had seen watching our arrival minutes beforehand had braved it in the pouring rain to the middle of the airfield where we were parked. I shut down, walked to a nearby portakabin and completed the aircraft documentation. The aircraft was fully serviceable. After this flight, I deserved a beer! We were driven to the operations complex where a small reception party greeted us. But first, we all had one thought in mind, to have a pee. I had a shower, and changed into a clean flying suit, before joining the rest for lunch. Before too long, all twelve aircraft had landed safely and not one had an unserviceability. By the time the second and third formations had arrived in the circuit, the weather had cleared up. The wind was still a bit gusty, but nothing like it had been an hour or two earlier.

An Andover had been arranged to fly us back to Binbrook. We thanked BAe for their generous hospitality, and made our way to the waiting aircraft. We were all very tired and were looking forward to getting home. After all, within a week we would be doing it all again, but that's another story, which I will save for another time!

Andy Williams safely on the ground at Warton [British Aerospace via Jimmy Dell]

ONE MAN, TWIN-FAN, ALL-ALUMINIUM PURSUIT SHIP

Chief Technician PHIL WALLIS
Chairman of the Lightning Association;
Formerly of XI Lightning Squadron

Anecdote One: **RAF Gütersloh, Mid-Seventies: Harrier v. Lightning**

Scene: Pilot bragging of the Lightning's climb rate to 20,000ft; Harrier pilot says he can do it quicker. Bet on: Both aircraft from start-up to 20,000ft, who is faster? Pilots walk, start up, Lightning taxies 1½ miles to runway threshold, Harrier is already at 10,000ft after doing a vertical takeoff! The Harrier won! This led to a famous cartoon circulated around Germany and UK:

Anecdote Two: **GAF Jever, circa 1979: F.15 v. Lightning**

Tactical Leadership Programme (TLP) was a multi-national gathering of NATO air forces carried out at GAF Jever in northern Germany during the Eighties. Multi-aircraft sorties were carried out throughout a four week period with forces acting as aggressors or defenders in a "European Theatre of War" scenario.

Nearing the end of the four weeks, at one of the many social gatherings the inevitable argument arose between aircrew as to their respective aircraft's performance. In this case it was a USAF versus RAF debate:

"An F.15 can easily outclimb a Lightning in a post-takeoff vertical climb."
"Oh no it can't."
"Oh yes it can."
And so on. To prove once and for all who had the upper hand, the gauntlet was thrown down and duly picked up. During one of the final sorties a pairs takeoff with minimum stagger was planned with a USAF F.15 and a Lightning.

Behind the scenes though, all that beer talk about the Lightning's legendary performance was not exactly true, so XI Squadron would have to stack the odds slightly in its favour! Fortunately the Squadron had a T-bird on detachment which had a lower all-up weight than an F.6 and did not normally require to engage reheat on takeoff. More good news—they were cleared for flight without the ventral tank fitted! Minimum fuel load on board, rip out a few non-essential black boxes, etc. and you have yourself a Lightning 'XR3i', similar in performance to the Mk.1A's used at Binbrook on the TFF (Target Facilities Flight) in the early Seventies.

On the day, both aircraft crewed up and taxied out to the threshold. F.15 full power, brakes off and rolling. Seconds later, Lightning in full reheat and rolling. F.15 takeoff, gear travelling, into the climb followed shortly by our T.5. Quite a large crowd had assembled on the pan to witness the event as word spread.

British faces fell as the F.15 soared skywards. The Lightning though had taken a slightly different trajectory, opting not to bleed off energy and speed by sitting on its arse but rather to pull up steadily into the climb. Although the F.15 had got the upper hand, the Lightning was gaining momentum and at about 2000ft nosed past and overtook its rival—what a sight! Jubilation (and consternation) abounded.

The icing on the cake though came when the Lightning landed and the brake chute deployed to reveal a message painted on the rear face of the chute: "UP YOURS YANKS!" and a Union Jack. Needless to say, everyone enjoyed a few drinks that night recounting the day's events, which have passed into aviation folklore.

A selection of other cartoon gems which appeared in station magazines from Binbrook and Finningley

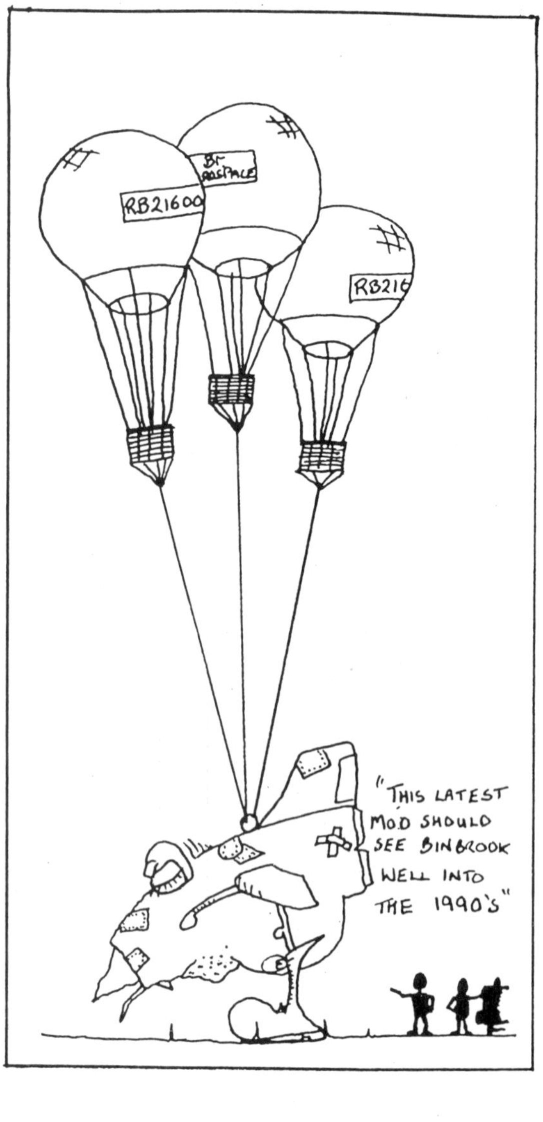

THE FINAL "AEROS" SEASON

Wing Commander JON FYNES

Formerly of LTF and 5 Lightning Squadron

When I was asked to be the 1987 Lightning display pilot my immediate reaction was one of delight. However, once I had realised that my only assets for the task were the last two Lightning F.3s, my own 1000 hrs Lightning experience and no idea of how to build a credible display, my initial enthusiasm was somewhat dampened. My first problem was deciding whether to plan a display that would please the several hundred thousand spectators that might see the display or the 35 Lightning pilots at Binbrook who would criticise me endlessly in the bar for not doing it "their way". Realising I could never impress the other pilots I decided to try and please the public! A keen display spectator myself, I knew that a boring display was either the excuse to rush off and buy a hot dog or to take some more photos of the static display. So I tried to build my own display to reduce the hot dog vendors' takings, believing that a fighter is expected to make a lot of noise and look powerful, not pretty. Another Lightning pilot gave me some good advice and suggested that "if the children scream when you start and let go of their balloons when you finish, then you've probably got it about right"!

So I planned the mandatory 6 minute display using reheat (for noise) as much as possible. In fact I used so much reheat that I had to commence my display from take-off, without holding, in order to be as full of fuel as possible. Similarly, I had to land immediately the display was complete as I was always on fuel minimums by the time I had finished. My calls of "fuel priority" when taxiing out for take-off often earned me ribald comments from the other display pilots, it also explains why I used to taxi rather faster than most. I did actually have to shut down and be towed back in at one display as the Red Arrows got in the way by blocking the taxiway after I had landed. The content of the display was left to me, although my Station Commander, Group Captain John Spencer, initially wasn't too happy about letting me do a barrel roll (especially as it was planned to include a 45° heading change) as barrel rolls were known to be one of the most difficult manoeuvres to fly and had been a killer. I was surprised just how difficult it was to design a display sequence that: spent equal time at either end of the display line; was capable of slipping from full show through rolling show to flat show in order to cope with varying weather; and even ensured that the slow-speed pass would always be into wind to make it look even slower. However, I finally ended up with a routine that everybody seemed reasonably content with and received clearance to " go public".

What about my two F.3's: XP741 (tail marking "AR") and XR716 ("AQ")? Both aircraft were destined for retirement at the end of the season and therefore they (unusually) had no airframe fatigue limitations beyond that of a maximum 6G. Despite my 1000hrs experience, I hadn't really appreciated the subtle variations between individual aircraft until I flew low-level aerobatics. For example, I always tried to display in "AR" because this aircraft was easier to slow-roll than "AQ". On the other hand, the engines on "AR" tended to protest rather more if I tried to throttle back the RPM on my final high speed pass to stop myself going supersonic. I was often asked by spectators why the F3 was only displayed with Firestreak missiles rather than Redtop; well Firestreak missiles actually generated a significant amount of lift which enhanced the handling and turning characteristics of the aircraft (it helped in combat too).

Jon Fynes taxies back at Binbrook in XP741 after a display practice, 1987 [Dave Drummond]

Of course many people were needed "behind the scenes" to ensure that my aircraft were always ready for a display and I think it only fair to mention my groundcrew here. Five groundcrew from 5(F) Squadron, led by chief tech "BJ", volunteered to give up nearly all their weekends for the summer (with no time off in lieu!) and travel to all my UK displays to give me the back-up I needed. They nearly always had to deploy to an airshow in a LandRover with trailer on journeys of up to 10 hours, and then often had to work into the night to sort out the aircraft after I had flown in. It was through their dedication that I didn't have to cancel a single display through unserviceability (unlike some much newer aircraft!). I hope those display organisers who, on one occasion, tried to give them mattresses in a corridor to sleep on, and on another occasion a tent, choke on their gin and tonics.

Jon Fynes rotates in XP741 'AR' during a display practice on 19th August 1987 prior to the Binbrook Airshow [HT]

Because 1987 was the last year for the Lightning display, I had 42 display venues rather than the normal 20-25. Some were at small town shows, while others like Weston Super Mare and Mablethorpe were along the beach—at Mablethorpe I actually did fly through a cloud of children's shiny foil balloons and for a moment regretted that Lightning pilot's advice! Most enjoyable though were the displays at airfields. There you could meet and socialise with the other display crews and see the crowds enjoying themselves. It was also at these displays that I fully came to realise the dedication of Lightning supporters. I kept bumping into or seeing the same spectators all over the country, many of whom specifically followed the Lightning display and had done for years. In particular, I remember talking to three men in their early twenties on two consecutive days at displays 150 miles apart. We chatted for a while and I then offered them the chance to sit in "AR" and get the "feel" for it. They refused, and when asked why replied," to sit in a Lightning has always been a dream, and to actually do it would end the dream for ever". I understand they finally got in an aircraft at the Binbrook open day.

Of course the highlight of the year had to be the last-ever Binbrook Air Show. At previous Binbrook shows we had seen a maximum of about 24,000 people and thought that for the last show we might get 30,000. On the day we had to close the gates at 45,000, even my brother couldn't get in! The weather was appalling, it drizzled all day and we were really worried about letting the crowds down. The Station Commander ordered me to sit in my cockpit and whenever the crowd got too restless he would scramble me to fly up and down the runway in reheat. As it turned out, although the rain never stopped the cloud lifted just enough for a reasonable show to be put on. I got to open the show and, because the Red Arrows had to cancel, to close it as well—just as the cloud lowered again. Despite the weather, it had been a marvellous day for all and it wasn't just the crowds who finished with a lump in the throat. A few displays later and my season was over. I said goodbye to "AR" and "AQ" and let them go into retirement. Without doubt, being given the opportunity to display the Lightning was the ultimate experience for me, there certainly will never be another fighter like it.

LTF disbandment, April 1987 - Dave Hart, Andy Holmes, Chris Allan, Clive Rowley, Jon Fynes, Ian Hollingworth & Dick Heath [via Jon Fynes]

DELIVERY OF XR728 TO BRUNTINGTHORPE

Flight Lieutenant CHRIS BERNERS-PRICE

Formerly of 5 and XI Lightning Squadrons

24th June 1988, a bright sunny day, and it's my turn to fly away from Binbrook one of the handful of Lightnings left in RAF service. The week before I had delivered my own aircraft XR753 (BP) to become gate guardian for the newly-formed XI (Tornado) Sqn at RAF Leeming. The RAF thought XR753 was going to RAF Coltishall to be burnt as fire-practice but somehow on its last trip it landed at Leeming—but that's another story!

The Lightning Preservation Group had bought XR728, personal mount of the Station Commander Gp Capt John Spencer and thus coded 'JS'. The Group wanted the aircraft delivered to a disused airfield called Upper Bruntingthorpe, near Leicester. None of us at Binbrook had ever heard of it, and we thought we knew every strip of concrete in the UK that you could put a Lightning down on. A Lightning needs 9000 ft of runway (a mile and a half) to stop if the brake 'chute fails, and at all times we need to know instinctively where the nearest safe strip is because of our very limited fuel. In the event, Bruntingthorpe turns out to have the longest useable runway in the country, having been used by the USAF to fly B.47's and RB.66's in the early 1960's before they discovered the aircraft were interfering with the local VLF transmitter for their nuclear submarines. Since then it has been used as a high-speed car test track and so the runway is still in useable condition.

Having spoken to the new owners on the 'phone to find out more details about the airfield and where they want me to park, as well as how I am going to make contact with them on the VHF radio, it is time to spend 10 minutes putting a few lines on a map as I want to fly there via the low flying system. Then I must fill in the Authorisation sheets to make the trip legal and put on the layers of clothing needed to fly a fighter aircraft.

Chris B-P walks out to take 'JS' for her final flight.
[Jim Featherby]

Firstly the T-shirt to cover the neck in case of fire or windblast, followed by the fire-resistant flying suit and thick leather boots to prevent injury on ejection; then the anti-G suit to prevent blood pooling in the legs and starving the brain of oxygen when turning the aircraft hard; next the sweat-resistant gloves and the individually-fitted 'bone-dome' that clamps tightly on to my head and which can weigh the same as a full suitcase in the middle of a fight under 6G. Over the Nomex flying suit goes the blast-resistant life-jacket with its emergency radio and flares, while around the lower legs strap the blue leg-restraints which will pull them hard against the ejection seat and prevent them being snapped by windblast in the event I have to 'go for the ride'. Luckily it isn't winter, otherwise thermal long-johns and the very heavy rubber-sealed immersion suit would also have to be added to the list. Wearing all that kit in a severely overheating cockpit while straining against the effects of 'G' means there aren't many fat Lightning pilots!

I've not found any signs of leaks or other problems while walking around the aircraft and so I climb up the ladder into the 2-storey high cockpit—the Lightning is a massive size for only one pilot. I remember one French Mirage pilot who was walking with me across the tarmac at Cambrai, having been 'shot-down' by a Lightning. He looked across from his clean, sleek, dainty little fighter to the slab-sided, dirty and leaking Lightning and said "C'est l'avion pour l' homme, n'est pas!"

Chris B-P straps in to XR728 'JS' which stands isolated on the flight line at Binbrook. [Jim Featherby]

While strapping in with help from the groundcrew, I scan the cockpit and put the switches into their correct places before asking for ground power to be connected. The Lightning needs a huge diesel ground power unit or 'Houchin' to provide 28v DC for the starter system and 115v AC for the instruments and radar. Once connected and with everything switched on, it's time to go for the start. Starting a Lightning can be a tense time because it uses 'Avpin' (IPN, Isopropyl Nitrate) as fuel for the engine starter units, one of which is bolted to the front of each RR Avon. Avpin is highly corrosive, poisonous and semi-explosive, it makes its own oxygen so once it catches fire it can't be put out—you either start up or blow up!

JS more-or-less behaves herself and I taxi out to take-off on the north-easterly Runway 03, gently dabbing the brakes to keep the aircraft from going too fast. On the ground at minimum power I already have the thrust of a Jet Provost and a Hawk pushing me along! On the runway I put the power up to 92% to check the brakes for slip, put on full cold power as I release the brakes and then rock the throttles outboard. The double kick of the reheats pushes me back into the seat—truly a fantastic feeling as 60,000 horsepower is unleashed for just one person. In a few seconds I am airborne with the gear travelling, and at 180 knots I snap 3 G and rotate the aircraft into the vertical as only a Lightning can.

As I'm going low level to Bruntingthorpe I roll off the top and let her accelerate as I descend back to 250ft, and boy does she go! The books say the Lightning can accelerate from undercarriage limiting speed to 90% of the speed of sound (Mach 0.9, about 650kts at low level) in 18 seconds! Most military aircraft will fly low level at 360kts or 420kts, but using only cruise power the Lightning can fly at 540kts (9 miles per minute). At that sort of speed England becomes a blur on either side and I have to search ahead for pylons, and light aircraft doing only an eighth of my speed. Being in a Lightning also brings certain responsibilities in that if one sees another military aircraft it has to be 'shot-down', while to be shot-down oneself is a cardinal sin. I also have to make sure I don't go supersonic, it would only take a few seconds of reheat at high speed and I would leave a sonic boom that would break hundreds of windows. One Lightning sonic boom over Anglesey blew down a barn!

As I get nearer to Bruntingthorpe I slow down and pull up in order to get a good look at this disused airfield and to make contact with the new owners of JS. I need to get radio contact with them to ensure that the truck race that has been going on has finished and to find out what the wind and pressure settings are. Landing a Lightning is probably the most dangerous part of any trip because of its relatively poor brakes, slim high-pressure tyres and long brake-chute. Today 'Air Traffic Control' is being provided by someone sitting in a Cessna on the ground.

My plan to land on the south-westerly runway is foiled by a strong tailwind, which means that I have to approach the north-easterly runway over a huge bank erected to shield the local village from the noise on the track. Before landing though, I must reduce weight by burning down to minimum fuel which is 800 pounds a side, or enough to hop to the nearest airfield in a straight line. Even though the aircraft is destined to stay at Bruntingthorpe, I have to have a diversion in case something is blocking the runway or the wind is outside limits.

The RAF is very twitchy about unauthorised air displays and I have been briefed very strongly not to do one. However, I have also been told to shoo any vehicles off the runway and to make a few practice approaches before landing. The Lightning does not have a fuel dump facility, and the only way to get rid of it is to put the fuel through the reheats. The RAF has charged the Lightning Preservation Group £1200 for delivery and so it seems a good idea to burn off fuel where they can see it. However, one lady obviously doesn't like the way I'm doing it as I can see her running away! (I think she only wants to look down a Lightning intake when it's safely on the ground.)

With 35 minutes on the clock it's time to land and I come in over the bank—180kts on the approach, 175kts over the threshold and touchdown, nosewheel on, pull brake-chute handle and because there isn't much fire-cover here and there is a risk of brake fire if the 'chute doesn't work, I shut down No. 2 engine to reduce the residual thrust. However, today everything works and as I go past the line of Nikons (and one sheepish lady) there are huge grins everywhere. The RAF may not care what happens to the Lightnings, but here I know the LPG will keep the last single-seat British fighter in flying condition—a great piece of British aviation history saved for the future.

As I shut down the No. 1 engine and peace returns to Leicestershire, I'm met by a happy crowd of people and a speed-freak of another kind—Barry Sheene. Until I landed my three-wheeler, Barry had held the airfield speed record on a bike. He doesn't seem to mind that I've just beaten his record in a tricycle designed in the 1940's, and what seems to impress him most is that I'd done it while braking and that I had a spare 1300mph left for overtaking!

Having checked the runway for obstructions, Chris B-P brings 728 low over the tower at Bruntingthorpe prior to landing. [Steve Gensler]

WARTON FOXHUNTER TRIALS

KEITH HARTLEY

Experimental Test Pilot, British Aerospace Warton
Formerly of 56 and XI Lightning Squadrons

For many years the only Lightning left flying at Warton was XP693, the first production F.Mk.6, a converted F.Mk 3. Once the Lightning development programme had finished in the 1970's, and with the advent of the Tornado, XP693 was used sporadically as a high-speed chase aircraft, the Lightning's supersonic performance still being a comfortable match for the multi-role Tornado at high altitudes.

XP693 was briefly flown in the markings of 65 Squadron to commemorate the Squadron's 75th anniversary. [Geoff Lee/BAe}

The Tornado F.3 was developed in the early 1980's from the basic GR.1 Tornado by extending the fuselage and fitting a significantly different avionics system, including a new AI.24 Foxhunter pulse-doppler interceptor radar. The development programme for the Tornado F.3 involved a considerable amount of flying to test and improve both the radar itself and the rest of the weapons system. The threats facing Tornado meant it was essential that it was tested against representative fighter targets, with high speeds, rapid rates of climb and good turning performance: the Lightning, being available and (relatively) cheap to run, was an obvious choice to fill this targetting need. Having said that, a careful study was first carried out of the various available potential target aircraft. The Jaguar was ruled out because its supersonic performance was too inferior, as was its turn rate. The Phantom had good supersonic performance, but the logistical problems of operating this (new to Warton) type meant it was far from practical—besides, at the time the RAF could not spare them for the task. Concorde was considered (a development Concorde was then still flying at Filton), but although its supersonic performance was outstanding, its size and manoeuverability meant it was too unrepresentative of the potential threat. Naturally therefore, the Lightning filled the bill perfectly: although the Tornado is a match for it in acceleration at low altitudes, the Lightning was still superior at medium and high altitudes, thanks to the high-altitude efficiency of the Avon turbojet in comparison with the Tornado's by-pass turbofans. However, the same difference resulted in significantly better range and endurance from the Tornado, the mighty Avon being decidedly thirsty in comparison when operating at lower altitudes.

To provide the necessary numbers of targets and sorties, the RAF allocated 4 Lightnings to Warton from Binbrook, when the RAF withdrew the aircraft from front-line service: XR724, XS904, XS928 and XR773, the latter going into store at Boscombe Down and joining the programme later when XR724 ran out of fatigue-life. All were relatively high time and high fatigue-used aircraft but still with ample left for the projected task. When delivered they all retained weapon packs (but no missiles) and serviceable radars. Initially they were operated without the overwing tanks, but these were soon fitted to give them more time on task and to better match the Tornado's sortie times. This of course limited them to subsonic

XP693 and her three sister ships at Warton in October 1992. [HT]

The final four Warton Lightnings line up for the last time on 22nd December 1992. (L-R XS904, XP693, XS928 and XR773)
(Below) The faithful ground crew [both HT]

speeds, but the gain in sortie time—particularly when carefully flown—gave good flight times, particularly when the target requirements included little hard manoeuvering, and thus avoided use of reheat. In addition to the Lightnings, Buccaneer XN974 was also provided as a target and had the added advantage that when fitted with bomb-bay and bomb-door fuel tanks, and a buddy-buddy refuelling pod, it could provide additional fuel (and time on task) for both the Lightnings and Tornadoes.

Many of the sorties took place over the Aberporth test range in Cardigan Bay, where fighter and targets were all tracked with precision ground tracking radars. The tracking data added to the huge amount of Tornado instrumentation data, all analysed post-flight by BAe's flight test engineers, from which they could accurately monitor the actual relative positions of the aircraft and compare them with the data derived from the Tornado's radar and weapons system. To enhance the ground tracking performance a special radar transponder was fitted to the target aircraft to improve the tracking radar accuracy, resulting in the addition of an extra cockpit switch to turn it on and off.

A typical sortie would begin with a detailed briefing between the flight test engineers and all the involved aircrew, covering in detail all the intercept geometries and parameters, the required manoeuvering, radio calls, test priorities, fallback plans in the event of unserviceability or bad weather, and all the myriad of detail that needs to be addressed if the maximum amount of data is to be achieved. The aircrew would then brief their "domestics"—weather, failure plans, formation procedures and recovery plans—and would then get kitted out and go to the aircraft. In the test environment it takes about 20 - 30 minutes to get a Tornado airborne from walking out, the usual operational procedures being additionally complicated by the extra instrumentation and test systems fitted to the aircraft. Because the Tornado is the critical part of the sortie—if the Tornado is u/s there is little point in anyone else getting airborne—it was normal for the Tornado to start, taxy and takeoff first, confirming serviceability when airborne, whereupon the Lightning targets would start up and follow. Equally normally, the Tornado would cruise at its relatively slow optimum cruise speed to the northern end of the test range, while the Lightnings would use their rapid climb performance to go to higher altitude and then cruise fast (0.9 Mach) to the southern end of the range. The geometry is such that the aircraft would generally all arrive in the right place at the same time, thus being ready to start with minimum fuel burn.

Thereafter, the interception geometries were usually set up by Aberporth radar, with variations and adjustments, if required, controlled by the Tornado. Target profiles varied widely, from high speed at very low altitudes to high supersonic speeds at high altitude. When supersonic targets were required the overwing tanks were removed, and the Lightning reverted to its normal GT performance—always popular, and well fought for by the core of Warton's Lightning pilots! The fastest profiles took considerable distances to set up because of the severe constraints preventing flight above the speed of sound overland, or even close to land. Typically, a Lightning target would start its supersonic acceleration to the east of the Isle of Man, heading southerly to pass clear to the west of Anglesey before turning into the range area proper. The Tornado, meanwhile, would accelerate from north of the coast at the south end of Cardigan Bay, again heading to pass west of Anglesey: the interception would occur near Anglesey, with combined closing speeds of well over 2000 mph. Such profiles are ideally suited to the Lightning, the whole design of the aircraft being optimised for supersonic acceleration at high altitude: the writer recalls cruising down the Irish Sea one starlit night at 1.8 Mach in XP693, without using reheat—the 1990s "supercruise" concept being demonstrated by a design already more than 30 years old!

Another pleasure for the Warton pilots was the chance to fly a single-seater with a radar. The ex-Binbrook aircraft had functional radars on arrival and although it was accepted that if they subsequently failed they would not be repaired, they stayed remarkably serviceable. Until the later advent of the single-seat Hawk with its F-16 radar, the Lightning provided the only opportunity for the pilots to demonstrate their radar handling skills against those of the Tornado navigators—a source of much lively banter in the crewroom! Unfortunately, the test requirements soon put paid to the Lightning radars. Looking at the Tornado radar instrumentation, the changes in the radar reflectivity of the Lightning as the scanner beat were quite apparent; naturally enough, the Lightning scanner is a good radar reflector, and reflects the Tornado's radar emissions very efficiently—and this proved to be an additional complication in determining the Tornado radar's performance. It was therefore decided to inhibit the scanners in order to give the Lightnings a more regular and quantifiable radar signature.

The requirement for more accurate postion data led to the modification of two of the aircraft (XP693 and XR773) with inertial navigation platforms. These were standard RAF Jaguar inertial platforms, complete with control panels, made by Ferranti. The new equipment was fitted in the weapons pack, there were virtually no controls in the cockpit and none of the navigational data was displayed—instead, all the required data was recorded on magnetic tape in the weapons pack. Warton's ex-RAF Lightning pilots, all of whom had spent many hours over the North Sea with no navigational aids at all (being well outside TACAN range and frequently below ground radar coverage) smiled wryly at the arrangement—a modern, self-contained navigation system fitted at last, but inaccessible to the pilot!

Finally, the aircraft were flown to benign speed and manoeuvre limitations. Although the Company aircraft, XP693, was flown for many years to within the company-cleared limitations (for example, speed limitations of 700 knots/2.1 Mach), with the arrival of the ex-Binbrook aircraft the general operating limitations were reduced to 90% of the RAF Release-to-Service limits. This meant that although many of the sorties required manoeuvering targets, the consumption of fatigue-life remained relatively low. This, coupled with the relatively low flying rate—typical in flight development, where time is required between sorties for data analysis and for modifying test equipment—meant that the aircraft lasted for considerably longer than they would have had they remained in RAF service. When finally disposed of by MoD at the end of 1992, only XS904 was fully fatigue-expired. The disposal of the aircraft brought to an end the UK's military life of the Lightning: it was perhaps appropriate that its last task in Service was to contribute to the development of the RAF's 1990's fighter.

Warton 23rd December 1992: Keith Hartley signs out XP693 for the delivery flight to the Lightning Management Group/Lightning Flying Club at Exeter. [Both HT]

ON THE OUTSIDE LOOKING IN

ANDY RICKARD

Lightning Preservation Group

Monday 6th June 1988, a day I remember well: my final 'sortie' to RAF Binbrook, the Lightning's last home. Quite some day though, with some of the most exhilarating and spirited flying I had ever seen from the resident aircraft. The sight and sound of XV328 'BZ' streaking over the airfield after a Mirage, or XS899 'BL' in a tail-chase with an F-16 will stay with me forever.

On reflection, many years later, a summer holiday at this point was bad planning as I missed the final delivery flights. I picked up a couple of aviation magazines in the Luton Airport departure lounge: every enthusiast will remember the covers all bidding farewell to the Lightning, with the Binbrook 'flagship' XR728 'JS' bowing-out in great style on many of the cover features.

The Lightning Preservation Group in its original form held its collective breath at this time, would we be successful in our bid for an F.6? It was agonising to be many miles away in Cyprus. After a week's holiday had passed my magazines were now decidedly 'dog-eared', I even killed time by driving to RAF Akrotiri to photograph XS929 being prepared for gate-guard duty. Mid-way point, time to ring up Chris Norris for an update. He tells me we've been successful, we've got an F.6, XR728.... The magazines were now studied in disbelief, no—surely not? XR728 was safely down at Bruntingthorpe by the time I returned, many things had changed in a short space of time. The planned spares-recovery of the aircraft had been cancelled leaving us with a fully active aircraft to care for. The rest as they say is history. Needless to say we all had a pleasant distraction from the end of Lightning operations at Binbrook, and many decisions to make regarding the future of 'JS'.

The thrill of getting XR728 has for me never faded, but by the spring of 1989 I was looking at road maps, where on earth was this place Warton? Blimey, must be 200 miles at least, maybe I'll try it just once..... Tuesday 4th July, and off we go, 6am start, a bit different to Binbrook departures. Into Freckleton by 9.30am, through into Warton with the airfield and hangars off to my left. Maybe left here....no, that's the main gate, security not amused, what about down here?.... Bank Lane. Round a few tight bends, and to my left out on the north-side flightline sat XP693 and XS904, looking great out in the summer sunshine. XR724 was thundering away on the test-pan side of the airfield, "Sunset Strip" I later found was the name given by Aerospace employees. XP693 took off in an easterly direction (Runway 08), great to see one airborne again after a year's lay off. 693 flew over several times at high level, and after 45 minutes or so joined the visual circuit with a Tornado and a Hawk. After three overshoots she landed, taxied into the flightline, and shut down. The activity around the two Lightnings suggested that was that for the day. Hmmm, worse than a bad day at Binbrook, but by the 11th I was back again. No Lightnings flew from Warton that day, but I was lucky enough to see XR773 fly in from Boscombe, still wearing its XI Squadron markings and BR tailcode. Wednesday 19th saw 693 and 904 flying out in a westerly direction (Runway 26) at 1pm, another lovely summer's day too. My slide and photograph collection was again on the increase, the bug by now had well and truly bitten. The significance of some sorties was a little unclear to me in the early days, ie. "shakedowns", C/T's, chase, etc., but by early 1990 I had a contact at Warton and it was nice to know when flying was intended. Many of my trips from this point on were made with LPG colleague Chris Norris, which diluted the expense somewhat and also lessened the boredom and disappointment of the "blank" days.

A rare chance to catch a Lightning elsewhere than Warton came on Monday 30th April 1990. RAF Waddington had held a public 'photo call' over the weekend and a guest from Warton, XS928, was due to depart sometime in the morning. She was in the dispersal close to the roadside with many other 'classics'—what a line-up: Phantoms, Hunter, Starfighter, Drakens.... XS928 started and taxied out to the Runway 21 hold, just yards from the A15 roadside. 'Tarnish 9' held there for some time (costing me a fortune in film) before getting airborne at 11.30, circling the airfield and then running through hard and low, obviously aiming for the BAe groundcrew on the dispersal. She shot over our heads so low that it was hard not to duck and then, by tradition, lit the reheat and stood on her tail. "Waddington departures, Tarnish 9, Binbrook request a fly-through if possible." "Tarnish 9, I don't see why not." Ever wanted to be somewhere else?

Friday 27th July 1990 saw me underway at 4.30am, destination Warton (where else?). XR724 was now said to be close to the fatigue limit and was flying to Shawbury for storage, planned departure 8am. She lifted off Runway 08 at around 8.30am with Peter Gordon-Johnson at the controls; hazy sunshine and very misty conditions failed to inhibit a lively farewell, 'Tarnish 6' also obliged with fly-throughs at Samlesbury and RAF Valley. So the Warton strength was abruptly reduced to 4 with XP693, XR773, XS904 & XS928 remaining on the radar-development programme.

A highlight of 1990 has to be Saturday 15th September, Warton Families Day. XP693, the 'silver bird' was on the flying programme, with XS904 in reserve. Thankfully it was warm and dry on the day. The previous Thursday I'd seen Peter Gordon-Johnson practising his routine, putting XR773 through its paces, so I knew what to expect. The morning consisted of a tour of Warton: XR773 & XS928 looked pristine in the 'Lightning' hangar—very frustrating as the usual policy of 'No Photography' still applied. Over on the south-side flightline XP693 & XS904 sat side-by-side in the sunshine. 693 was 'clean' with the overwing tanks removed while 904 was still fitted with these ungainly additions. Thankfully 693 started up faultlessly and taxied out for departure, by this time I was outside the airfield, I couldn't let this pass without a photograph! 693 got airborne and promptly flew off to Samlesbury, unknown to me she was due to fly in their programme too. It was worth the wait when she returned to the Warton circuit: Peter Gordon-Johnson always seemed to fly the Lightning with so much enthusiasm and today was no exception. High angle-of-attack slow flypasts, hard turns pouring vapour from the wingtips, and of course reheat climbouts were all performed with his usual vigour. It was great to see one flown like this again as too often when watching the radar flights the Lightnings were forced to land from a single approach, 'fuel priority' after being airborne for anything up to 1½ hours.

Peter Gordon-Johnson puts XP693 through its paces at the Warton Families Day, 15th September 1990. [Andy Rickard]

Lightning activity varied greatly depending on the current programme and aircraft serviceability. The aircraft were rotated through the various services and checks, the flying workload spread carefully to maximise airframe life—particularly with the ex-Binbrook machines, and so it was very rare to learn of more than two being serviceable at any one time. It was always good to catch a Lightning coming off a deep service as they would be put up on air-test, or 'shakedown' as it was known locally. Any snags or adjustments were then dealt with. 'Shakedowns' from my point of view were worth maximum effort for a trip up north as they often involved a great deal of circuit work, with up to 6 or 7 overshoots and some spirited flying into the bargain.

Wednesday 7th November 1990 saw three of the Lightnings serviceable and due to fly on radar sorties. I travelled up with high expectations, but prepared as usual for disappointment. Many planned sorties were 'scrubbed' due to weather, an unserviceable Tornado or reasons very unclear. Today it all fell nicely into place. XR773 & XS928 thundered off Runway 08 together, simultaneously tucking up the gear and then banking hard right, the reheats glowing nicely against an overcast sky. The Lightnings could be seen accelerating behind the southside hangers, climbing through the high cloudbase and then rumbling away in the distance for long after. At around midday they reappeared, having been airborne for 1¼ hours.

XR773 lines up on Runway 08 at Warton at 10am on Thursday 28th March 1991 flown by Peter Orme ('Tarnish 3')
[Andy Rickard]

XS928 turns prior to departure on Runway 08, Wednesday 29th May 1991, for a 'shakedown' flight (air test) after servicing. Flown by Peter Orme.
[Andy Rickard]

XS904 turns after dropping the brake 'chute at Warton on Monday 21st September 1992.
[Andy Rickard]

Both aircraft obliged with single overshoots whilst in the circuit, my film was complete as 928 turned and dropped her chute opposite my vantage point. By late afternoon I was sure that the single pm sortie would be cancelled but at 4pm 693 proved me wrong. She too departed from Runway 08, tucked her gear up very early and turned hard right, her reheat glowing furiously in the fading light. She passed over the hangarage to the south, tip-trailing in the damp air; a down-side to this time of year meant 693 landing in pitch blackness.

I had to wait until Tuesday 22nd January 1991 for the next 'double'. XS904 & XR773 were due to fly on a radar development sortie, but when I arrived at 10:15 only 904 was out on the line. The given time of 11am came and went with no sign of activity or of 773. At 11.30 the groundcrew made their way towards 904; I always found this a little nerve-racking as they sometimes fitted the towbar, which signalled the end of proceedings. However, off came the intake blank and the Houchin power generator started and howled away, all looked promising. 'Purple One' arrived with Peter Orme on board ('Purple One' being the pilots' name for the Sherpa crewbus) and after the walk-round checks 'Tarnish 3' climbed aboard. 904 started faultlessly, taxied out and was quickly airborne from Runway 08. From the R/T exchanges it was obviously a crew-training sortie during which 'Tarnish 3' approached RAF Valley and RAF Brawdy on practice diversions before returning to Warton an hour later. Before landing Peter overshot four times, 904's camouflage standing out smartly against a clear blue winter sky. Groundcrew activity suggested she might be flying once more, with refuelling and chute fitting being carried out. At 2.45, sure enough 'Purple One' pulled up alongside 904 and by 3pm 'Tarnish 5', Derek Reeh, was shattering the peace, getting airborne from Runway 08. 904 rejoined the circuit at 4pm and like Peter Orme he carried out several overshoots. Not quite the double I was expecting, in fact somewhat better with more circuit work than I could have expected from the scheduled radar sortie. A lovely winter's day too: still, frosty and sunny.

Tuesday 19th March 1991. XS928 was due to fly on a C/T sortie in the morning, with 693 and 904 due to fly in the early afternoon on a chase sortie. Following 693's successful 'shakedown' on 28th February, Warton now had all four Lightnings serviceable and XR773 was back-up for the day's flying programme. Driving up Bank Lane early that morning I was greeted with a stunning sight, all four Lightnings lined up on Warton's short-runway! At 9.45 the activity around 928 suggested my information was spot-on. By 10am she was thundering off Runway 26, the call-sign 'Tarnish 6' identifying Peter Gordon-Johnson at the controls. 928 turned left on departure to circle the airfield, staying low and accelerating all the time, running through low and hard before pulling-up for the traditional departure—good 'Tarnish 6' stuff! The break into the Warton circuit some 50 mins later equalled the excitement of the departure—a hard low run down the runway, pulling up steeply over the tower and then a Derry turn, a ¾ roll to the right leading to a left exit into the circuit. Peter Gordon-Johnson followed up with several varying approaches. The C/T sorties always seemed to include single engine and flapless overshoots, familiarizing the pilots with the differing handling characteristics in the event of a failure. After landing, 928 taxied into the line-up on the short runway and shut down—being towed away shortly after, her task complete for the day. The pm 'double' with 693 & 904 was cancelled for some reason, but 773 saved the day by going up later on a second C/T sortie. The usual North-West weather had by now closed in, away with the camera I thought, stand and watch, soak it all up.

Friday 7th June presented another opportunity to catch a Lightning away from Warton. This was always welcomed as it allowed you to relax, camera in-hand away from Warton's strict security. 693 was visiting RAF Wattisham for the base families day. A credit to Warton she was too, shining brightly in her natural metal finish and complete with 56 Squadron markings. The red and white checked nose board and phoenix on the tail looked absolutely stunning. Peter Orme piloted 693 to Wattisham that day and after a lively arrival he taxied in, parking the star of the show outside its allocated HAS. I'm sure Peter felt like a film star as a sizeable crowd gathered around the aircraft, such is the impact of the Lightning. It was nice to walk around the aircraft, to inspect her at close quarters rather than to peer through binoculars from a distance. Derek Reeh delivered her safely back to Lancashire on the Monday at around 5.45pm, needless to say I had travelled up to record the event with my trusty Minolta.

1991 continued with the Lightnings remaining fairly active, a good mixture of 'doubles', C/T's and 'shakedowns'. As well as flying targets for resident Tornados AS54 & A03, the Lightnings were also working with RAE Bedford's Buccaneer. I had encountered a high failure rate in the latter part of 1991 as this particular aircraft seemed very prone to going u/s, at least that's the explanation I often received!

On Friday 11th October 1991, a light easterly wind saw XS904 about to depart from Runway 08, looking pristine as she turned to line up in the Autumn sunshine. Derek Reeh ('Tarnish 5') was to fly XS904 to RAF Wildenrath in Germany to mark the closure of the base. This was the final overseas visit by a Lightning, and when Peter Gordon-Johnson flew her back on the 14th XS904 had been 'zapped' on the fin with a 19 Squadron dolphin insignia, which she wore until 1995 (when she was returned to XI Squadron markings).

(Above) XS904 piloted by Derek Reeh arriving at Wildenrath for the final overseas visit by a Lightning on 11th October 1991, and (Below) a dramatic departure on the 14th, flown by Peter Gordon-Johnson. [both Karl Kähler]

As 1991 drew to a close, talk of 904 nearing the end was on the increase as she had well exceeded 4000hrs, although as history now shows she survived until the end. By the spring of 1992, 928 was withdrawn for major servicing during which a crack was detected in one of the fuselage formers. All four Lightnings were grounded and their future on the programme hung in the balance for several weeks. Thankfully the green light to proceed was eventually given, the Lightnings returned to work and a repair for 928 was undertaken. Sadly she was never to fly again. By the latter part of 1992 the Lightnings were again under threat. Too many experienced Lightning personnel were leaving the company and as a result the Lightning's participation in the programme was ended on 30 November 1992. 928's fate was sealed long before that date, but 904 was spared to assist XP693 & XR773.

1992 was not all doom and gloom, apart from the period of grounding the Lightnings remained very active throughout the year. XS904 visited Boscombe Down on Friday 12th June for the airshow, returning to Warton on the Monday. Two days holiday were booked as I had to attend both days. Monday's departure by 904 was worth the wait (she was the last aircraft to leave). Derek Reeh, 'Tarnish 5', held 904 close to Boscombe's long runway and climbed rapidly to 21,000ft, she was visible for the climb on that hot clear day. On Tuesday 16th June, 693 took to the air at Warton to celebrate 30 years since her first flight. Peter Orme flew a lovely smooth display routine on a warm, bright summer's day. Monday 29th June and a trip to BAe Woodford was the order of the day. XP693 had been on display at the families day and was to return to Warton: Peter Gordon-Johnson departed vertically at around 1pm and again the weather was very kind. There then followed a mad dash to Warton as 'Tarnish 6' was due to take 904 up on a radar sortie with A03 at around 2pm. I arrived at 2:20 as she was just about to taxy out, the Warton 'slip' for once working in my favour. 904 departed from Runway 26: Peter Gordon-Johnson kept 904 down low, building up speed, then pulled up in a near vertical climb, rolling all the way, up through the high cloudbase. How can you call that work? A fast run and break back into the Warton circuit stirred the air for long after he had passed. Several overshoots followed, tucking the gear up and lighting the reheat. Although extremely tiring, it was very pleasing to see two different Lightnings in differing locations on the same day. XP693 returned to visit RAF Wattisham on Friday 3rd July 1992, again in 56 Squadron colours, to mark the squadron's disbandment over the weekend. The weather was not so kind on this occasion. Derelict Phantoms in the graveyard were very reminiscent of the same butchery that had taken place at Binbrook some four years previously. I was camped out at Warton on the Monday for 693's return and had the added bonus of 904 flying with A03 at 2.15pm.

The ultimate trip to Warton for me was to take place on Thursday 13th August 1992. I had permission on this occasion to photograph from the pan a double sortie with 904 & 693. The transport took us over to the south side hangars, 904 & 693 sat outside the hangars hooked up to the Houchins and ready to go. The feeling of being able to photograph freely at Warton was tremendous. 'Purple One' delivered the pilots to their aircraft at 9.50am, Peter Gordon-Johnson to XS904 and Peter Orme to XP693. Soon the noise from the Houchins added to the atmosphere, four long hard blasts from the Avpin-fired starters brought the Avon engines to life and after the usual short delay for pre-flight checks the two aircraft taxied from the pan. 'Tarnish 3 & 6' departed together from Runway 26, cycling the undercarriage in unison and thundering off in a westerly direction. I photographed XR773 on servicing in the hangar whilst waiting for 904 & 693 to return. Also photographed was 928, tucked away in the hangar corner looking a little forlorn.

XS904 has the intake blank fitted after flying in a 'double' with XP693 on Thursday 13th August 1992.
[Andy Rickard]

693 & 904 broke back into the circuit at around 11.45, both Lightnings banking steeply overhead and turning downwind for finals. Peter Orme landed 693 from the first approach, but Peter Gordon-Johnson had other ideas, overshooting twice and standing 904 on the re-heat. A hasty retreat to the pan was necessary to catch 904 & 693 taxying in. I certainly felt very privileged on that day—the photographs taken, in particular the pan shots, would not have been possible from outside the perimeter fencing.

Peter Gordon-Johnson departs Runway 26 in XP693, with distinctive 56 Squadron chequerboard nose markings, on 24th April 1991. [Andy Rickard]

Peter Gordon-Johnson's flying career at Warton came to an end on Friday 4th September 1992. His final fast-jet sortie was fittingly in a Lightning, Warton's 'silver bird' XP693. I was to witness surely one of the fastest flypasts ever made at Warton. Scarcely subsonic and with the upper surfaces of the Lightning completely enveloped in vapour as he pulled up hard in a near vertical climb over Runway 26. The climb, according to R/T chat, topped at 23,000ft, amazing to watch as he rolled 693 all the way up. Wednesday 23rd September 1992 and Keith Hartley stepped back into the 'hot-seat'. Keith was called upon to assist with the final Lightning sorties from Warton (I'm sure he hated every moment!) and a C/T was necessary to familiarize him with the aircraft again. 'Tarnish 9' took 904 up from runway 26. Even from a spectator's viewpoint all the Warton test pilots seemed to have their own style and Keith was no exception. He put 904 through its paces that day, the circuit work was tremendous, the overshoots very low, the turns tight and crisp and the run and breaks something to savour. Keith was doing a great deal of work with the LPG at the time (and still does) and so it was especially nice to have XR728 taxied by a *current* Lightning pilot.

The final Lightning three ship lines up on Runway 26 at Warton on Thursday 16th December 1992. (Front to rear) XP693, XR773 and XS904. [via Phil Wallis]

(Top) XR760 'H' of 23 Squadron returning to Leuchars; XR760 crashed into the North Sea on 17th July 1986, Bob Bees ejecting safely. [Rolls Royce plc] *(Centre) Saudi F.52's for the 'Magic Carpet' contract on the line at Warton. Front to rear: 52-657, 52-656, 52-658, 52-655* [British Aeropace]
(Bottom) Kuwaiti T.55 55-410 being held on the brakes prior to take-off from Warton

Plate 4

Flt Lt Peter Chapman of 2T Squadron, 226 OCU, and winner of the Embassy Aerobatic Trophy taxies in at Woodford in XP696 on 17th August 1974 trailing the red 'Aeros' brake parachute, with the rolling Cheshire countryside behind.

[HT]

(Left) The 5 Squadron final 9-ship formation on 7th November 1987

[Geoff Lee/BAe]

(Below) A beautiful shot of XS904 en route to Wildenrath on 11th October 1991

[Ian Black]

BINBROOK MEMORIES

T.5 XS420 'DV' (now being restored by Murray Flint at Fenland Aviation Museum) and XP749 'DB' return to Binbrook over the snow-covered Lincolnshire Wolds.

XS419 'DV' of the LTF carries a dayglo star in front of the fin flash during an exercise.

[Terry Senior]

XS901 'AH' pulls up using full reheat on 11th October 1986 during the 21st Anniversary celebration of 5 Squadron's equipment with Lightnings. [HT]

Mishaps

XS416 'AT' leaves Binbrook's Runway 03 after collapse of the starboard undercarriage leg on 19th July 1984. XS416 was repaired to fly again and after a spell as a derelict at Rossington is now preserved by the NATO Aircraft Museum at New Waltham, near Grimsby.

[all Terry Senior]

(Left) A light grey F.3 trails a half-open brake parachute around the Binbrook circuit before managing to jettison it and land safely. (Right) XS452 after taking the barrier, piloted by Simon Braithwaite. [both Terry Senior]

Two views of XR752 'H' of 5 Squadron after landing with the undercarriage up on 29th September 1977. The recommended procedure for undercarriage malfunctions was ejection due to the risk of the aircraft flipping on to its back during a belly landing. By 25th January 1978 XR752 was back in the air, but was the first F.6 to be withdrawn from use for spares recovery and was scrapped in 1987.

[Crown Copyright/MOD]

(Above) XS895 'BC' wins the Binbrook ploughing competition for Iain Small on 15th January 1986. On an icy runway with a crosswind gusting on the limits, the brake 'chute was taking Iain left towards the runway edge; upon jettisoning it the aircraft slid off the opposite side. [left, Terry Senior; right via Iain Small] *(Below left) F.6 XS903 of the LTF is recovered after nosewheel failure on 14th September 1979; a year later XS903 resumed its long career and in 1988 it became the black-finned XI Sqn flagship before being preserved at Elvington.* [MAP] *(Below right) XS457 'AT' on 9th December 1983 after it left the runway and the port undercarriage collapsed.* [Crown Copyright/MOD]

XS919 'BB' breaks into the circuit on an unusually sunny day at Binbrook. XS919 is now preserved at Devonport.

[Terry Senior]

XS459 'DW' displays smoking tyres on touchdown; it is currently preserved at the Fenland Aviation Museum, Wisbech.

[Terry Senior]

The fate of many Lightnings as Binbrook neared closure: here XM181 is attended to by the scrapman. The airframe was used to test a scheme of dark green camouflage in 1975 and then employed for fire rescue practice and as a decoy for many years.

[John Jackson]

LIGHTNING PRESERVATION

Presented below is a brief resume of a small number of preservation societies who, like the LPG, have the preservation/restoration of Lightnings as their sole or major preoccupation. Due to space limitations. this is perhaps a somewhat personal and incomplete list and there are a number of individuals, for example Charles Ross, Murray Flint and Wensley Haydon-Baillie who in an ideal world perhaps merit more than this passing mention.

LIGHTNING ASSOCIATION

(Top) XR724 en route to Binbrook from Shawbury on 23rd July 1992 [Tony Hulls]
(Bottom) XR724 after engine running at the 1995 Lightning Rally at Binbrook. [HT]

The Lightning Association was formed in 1988 after the Lightning's withdrawal from RAF service and has a membership scheme for those interested in Lightning preservation (the LPG and most other Lightning societies, with the exception of the Lightning Flying Club, do not have the manpower to cope with public membership). Apart from the grapevine and 'crash-gate rumour', the Association has until recently been the main source of information on all things Lightning through Charles Ross' editorship of the admirable Lightning Review. What was officially to be the last Review has now been published, but there are encouraging signs that Charles is beginning to cave in to the pressure for "just one more".

The Association purchased XR724 after its withdrawal from use at BAe Warton, thanks to a successful tender bid by Peter Talbot. XR724 was made airworthy after 18 months at RAF Shawbury by a team including Barry Pover, Phil Wallis, Colin Frise, Mick Cameron, Brian Lewis & Bas Livesey of BAe, plus Shawbury personnel and many others who loaned expertise or equipment. Most vital of all was BAe's agreement for Peter Gordon-Johnson to ferry XR724 home to Binbrook on 23rd July 1992 under their aegis, as the CAA would not allow a civilian operator to do so. It is intended to maintain the aircraft in serviceable condition, and engine runs were demonstrated at the 1995 Lightning Rally following the fitting of two 'new' Avons.

Contact: Lightning Association, Binbrook Airfield, Lincs. LN3 6BZ.

LIGHTNING FLYING CLUB

In 1989 Barry Pover and the late Vernon Piper began to gather together the nucleus of what was to become the Lightning Flying Club, the objective being to form a group to support the restoration of Lightning T.5 XS451 (G-LTNG) to flying condition at Plymouth Airport and ultimately to demonstrate the aircraft on the Airshow circuit. Before the rebuild was complete, and having obtained a large reserve of spares, the proprietors of 451 became also the proud owners of two airworthy F.6 Lightnings when the Warton trials aircraft were withdrawn in December 1992. XP693 (G-FSIX) and XR773 (G-OPIB) were delivered to Exeter Airport on 23rd December 1992 and Bas Livesey and the Warton apprentices spares-recovered the grounded XS928 at Warton before Barry Pover and Ernie Marshall donated the airframe to BAe for gate guard duties. Barry has recently changed the name of the maintenance and operational side of the LFC from the Lightning Management Group to the Classic Jet Aircraft Company (CJAC) and carried out work on Hunter and Canberra aircraft which has helped fund refurbishment of the hangar at Exeter, including provision of three-phase electrical supplies which will enable operation of hydraulic rigs and a power unit for the Lightnings. Increasingly large formations of Barry's Hunters are becoming a familiar and stirring sight at British airshows. The reluctance of the CAA to grant permits to test or to fly have so far kept Lightnings out of the air, but the team are hopeful that this could soon change. All Lightning fans must fervently hope that this is the case as CJAC is the only organisation in the UK capable of technically and financially keeping Lightnings flying; they now have CAA approval to work on and operate ex-military aircraft under BCAR A8-20 and CAP 632, and with CAA permission CJAC flight-tested ex-Boscombe Down Buccaneer S.2B XW988 (an aircraft type which Barry believes to be far more complex than the Lightning) and ferried it to Mike Beachy Head of Classic Jets, South Africa, departing the UK on 5th May 1996. The South African CAA has licensed Barry on the Lightning and it is intended that T.5 XS452, which was restored by Tony Hulls at Cranfield, will be given a Check 3 service by CJAC in preparation for a ferry flight to Sydenham, Belfast where it can be shipped to South Africa from the adjacent docks; this is a similar procedure to that in which 74 Squadron's T.5 XV329 was 'bagged-up' and delivered to Singapore in the late Sixties. (See also under T.5 Projects.) XS452 would join other classic British aircraft delivered there by CJAC, which includes F.6 and T.8C Hunters in addition to the Buccaneer.

(Top) XS451 under restoration at Plymouth City Airport in 1992.
(Bottom) XP693 with XR773 behind at Exeter in 1993.
[Both HT]

Contact: Lightning Flying Club, c/o Plymouth Executive Aviation, Plymouth City Airport, Crownhill, Plymouth, PL6 8BW.

LINCOLNSHIRE LIGHTNING PRESERVATION SOCIETY

The LLPS (John Jackson & Stewart Scott) became the first private group to acquire a Lightning for preservation on 20 December 1987, when Lightning F.3 XP706, having had its fin chopped and wingtips clipped, was towed from the Binbrook decoy line by tractor along narrow country lanes, via Caistor High Street, to Strubby Airfield. Lack of space in their hangarage has so far precluded the fitting of a new fin and wing(tip)s and the Society is *still* looking for suitable accommodation.

XP706 seen at Strubby in 1990. [John Jackson]

NATO AIRCRAFT MUSEUM
(alias GRIMSBY AIRCRAFT PRESERVATION GROUP)

(Top) XR770 displayed at New Waltham in 1994.
(Bottom) XS416 seen on 12th May 1996 after restoration.
[Chris Brydges via Trevor Garrod]

Prior to the closure of RAF Binbrook in 1988, and thanks to the efforts of Chris Brydges, the people of Grimsby were presented with Lightning XR770 by then Station Commander Gp Capt John Spencer on behalf of MOD. The aircraft was dismantled by RAF technicians and members of the Grimsby Aircraft Preservation Group and transported to Grimsby in July 1988, but there was much local outrage when Grimsby Borough Council declined to provide a location to display the aircraft.

With a view to attracting sponsorship, Chris decided to reassemble the aircraft at a temporary location near Laceby in the grounds of a local firm, and led by Phil Wallis the assembly work continued over a five-month period. Unfortunately, just prior to completion, opposition from a local parish councillor over planning permission caused the owner of the firm to request the group to vacate the site as soon as possible. The Lightning was dismantled again, transported, and then reassembled for (hopefully) the final time at its current location of Peaks Top Farm, New Waltham. The spurious 'JS' markings which XR770 wore on disposal were removed and the aircraft restored to its well-known colour scheme of red-fin, spine and leading edges adopted by then Wing Commander Andy Williams when XR770 was his 5 Sqn 'flagship'.

Phil Wallis and other group members then became involved in marketing the ten spares-recovered Lightning airframes on behalf of George Beck at Tanks & Vessels Industries (TVI), Rossington, near Doncaster. Some were sold as complete airframes (albeit with wings and fins removed and cockpits largely stripped) while others which had suffered damage or for which parts were unavailable were sold as cockpit sections. In return the group received T.5 XS416 to restore and display alongside XR770; Trevor Garrod's nose section of XR757 is also present which has cockpit lights working and ejection seat motoring up and down! Over 12 months of Sundays the group also restored XS897 'AC' which was originally planned to remain at Rossington. However, when George Beck left the firm the decision was reversed and major damage was done to 897 before a JCB was called off; the remains were then transferred to Firbeck for a second restoration attempt by South Yorkshire Aviation Museum. Of the ten original airframes at Rossington, XR725, XS416 & XS897 survived 'complete', XR726, XR747, XR757, XR759 & XS932 as nose sections, and XS419 and XS935 were unfortunately scrapped.

TONY HULLS' T.5 PROJECTS

Tony Hulls was a founder member of the LPG and was central to the maintenance of XR728 for a number of years. As it became increasingly unlikely that Arnold Glass would get CAA permission to fly any of his six Cranfield Lightnings, Tony began to take an interest in T.5 XS452 'BT' as it seemed to be the 'best of the bunch'. He took pains to keep the aircraft covered and Waxoyled and on two occasions he organised teams of willing volunteers to run the engines. In 1993 his patience was finally rewarded as he was able to buy the aircraft and, with a small team of helpers, begin to return it to serviceable condition. Incredibly, after six years of open air storage and inactivity, less than a year later 'BT' was fast-taxied at Cranfield Airshow to the delight of the large crowd. Further testament to the work of Tony and his team (Dave, Steve, Jon, John, Steve & Scott), headed up by Bas Livesey and pilot/engineer Liam Hughes, was the purchase of the aircraft by businessman Mike Beachy Head of Classic Jets in South Africa for it to be shipped to South Africa in late 1996 to join the growing numbers of classic British military aircraft there on the Air Show circuit. Plans are now in hand to get air under the wings again (see under Lightning Flying Club.) However, having restored one Lightning, Tony wanted another challenge and so scraped together the finance to buy the only other surviving Lightning at Cranfield, T.5 XS458 DY. XS458 appeared to be much more of a 'basket-case' than '452 and looked to be in poor condition. Nevertheless, within a few months an amazing amount had been achieved: both engines run; the undercarriage cycled with the aircraft on jacks; the air turbine gearbox replaced; the aircraft successfully fast-taxied in company with XS452; and now fast runs in reheat! XS458 has recently had a complete repaint in the unique LTF scheme which the aircraft previously wore for camouflage trials. Tony thinks that XS458 is now in a similar state of serviceability to '452 and with such boundless enthusiasm and sheer hard work, who knows what he and his team might yet achieve?

XS452 'BT' during a fast taxi run at Cranfield in 1995.
[Liam Hughes/Tony Hulls]

(Top) Bas Livesey and Tony Hulls remove the No. 1 reheat pipe from XS458 'DY' on 16th December 1995, prior to replacing the air turbine gearbox. [Geoff Wilkins] *(Bottom) A unique double taxi run by Lightning T.5's XS458 and XS452 at Cranfield on 17th February 1996, prior to the latter's sale to South African Mike Beachy Head.* [HT]

KEN WARD'S NORTH YORKSHIRE AIRCRAFT RECOVERY CENTRE

A one-man Lightning preservation team, in addition to restoring the cockpit of XM169 (the first F.1A), Ken won the Lightning Association 1995 'Spirit of Lightning Preservation' trophy for his dogged efforts against major obstacles to preserve a rare F.3, XR749, which he rescued from the Battle Damage Repair Flight at Leuchars. The airframe has now been restored to static display standard and returned to its unique LTF Anniversary blue fin markings at Teesside Airport, significant in Lightning history as RAF Middleton St George, former base of the Lightning OCU. An enormous amount of work was put in by Ken, at considerable expense to his health, first of all to prevent '749 being scrapped as it fell foul of planning regulations at its initial storage location, and then to organise the restoration work, ably assisted by Wally Douglas and volunteers from RAF Leeming. However, at long last the project has been completed; the aircraft was rolled out temporarily on 28th September 1995 and in May 1996 XR749 was placed on permanent display at the airport as a reminder of its military history.

(Left) XR749 in 56 Squadron commemorative markings on the line at Binbrook in 1985. [Dave Drummond]
(Right) Ken Ward and Wally Douglas during restoration of XR749 at Teesside Airport. [via Ken Ward]
(Bottom) XR749 after restoration, in front of the tower at the former Middleton St. George [Geoff Hill via Ken Ward]

LIGHTNING SURVIVORS

This is a somewhat optimistic list of extant Lightning airframes, as it includes some which have been shot-up on military ranges and others which are represented only as nose-sections. Inevitably, as time goes by most of these airframes (even those 'preserved') will gradually deteriorate and a number will be scrapped. It should also be emphasised that many of them are not generally available for public viewing as most are on private or MOD property.

Those in italics no longer exist but are included to show the fate of those which perished fairly recently.

KEY: <FF> = Front Fuselage, including cockpit
Cockpit = Cockpit area only survives
BDRT = Battle Damage Repair Training

Development & RAF Aircraft

WG760	P.1A	RAF Cosford Aerospace Museum, N. of Wolverhampton
WG763	P.1A	Museum of Science & Industry, Liverpool Road, Castlefield, Manchester
XA847	P.1B	Haydon-Baillie Museum. Stored Southampton (although exhibited at DRA Farnborough for SBAC Show 1994)
XG325	F.1/DB	Cockpit. 1476 Sqn ATC, Rayleigh, Essex
XG329	F.1/DB	Ian Hancock. Norfolk & Suffolk Aviation Museum, Flixton, nr Bungay, Suffolk
XG331	F.1/DB	<FF>.Gloucestershire Aviation Collection, Gloucestershire Airport (on loan from CARG)
XG337	F.1/DB	RAF Cosford Aerospace Museum, N. of Wolverhampton
XL629	T.4	DTEO Boscombe Down, nr Salisbury, Wilts. Gate guard
XM135	F.1	Imperial War Museum Duxford, Cambridgeshire
XM139	*F.1*	*Scrapped by 3/94. DTEO, Pendine Ranges, nr Tenby, Wales*
XM144	F.1	<FF>. SouthWest Aviation Heritage, Eaglescott, nr Ashreigny, Devon

F.1 XM144 after removal from the gate at Leuchars and before being reduced to a front fuselage
[John Jackson]

XM147	*F.1*	*Scrapped by 3/94. DTEO, Pendine Ranges, nr Tenby, Wales*
XM169	F.1A	<FF>. Ken Ward, N. Yorks Aircraft Recovery Centre, Chopgate, N. Yorks.
XM172	F.1A	RAF Coltishall, Norfolk. Gate guard. Owned by Haydon-Baillie Museum.
XM173	F.1A	RAF Bentley Priory. Gate guard—*up for tender?*
XM178	F.1A	Savigny-les-Beaune Museum, France
XM191	F.1A	<FF>. RAF Exhibition, Production & Transportation Unit, St Athan. (Described as F.6)
XM192	F.1A	Charles Ross. On loan to Bomber County Museum, Hemswell for restoration
XN726	F.2A	<FF>. Nigel Towler, Rayleigh, Essex
XN728	F.2A	Hulk by A1. A1 Commercial Vehicles, Balderton, Newark
XN730	F.2A	Luftwaffen Museum, Marseille-Karserne Airbase, nr Uetersen, NW of Hamburg

F.2A XN732 modified as a MiG look-alike at Laarbruch; sadly now scrapped.
[Keith Watson collection]

XN734 F.2A/F.3A Scrapped at Cranfield 1.12.94 (G27-239)
XN769 F.2 <FF>. Russell Carpenter, New Eltham
XN771 F.2A Scrapped DTEO Shoeburyness, late 1995

F.2A XN771 at Shoeburyness in August 1995, shortly before it was scrapped. Some of the engine accessories contain radio-active thorium which must be removed before scrapping.
[HT]

XN774 F.2A Burnt-out hulk, RAF Coningsby
XN776 F.2A Museum of Flight, East Fortune, nr A1, Lothian, Scotland
XN782 F.2A Luftfahrtausstellung in Hermeskeil Museum, Germany

F.2A XN782 at Flugausstelling Hermeskeil Museum, Germany in January 1995.
[Peter & Margaret Thompson]

XN784 F.2A Moenchengladbach Airport, Germany. Fuselage serial incorrectly reads 'XN781'

XN795	F.2A	<FF>. Nigel Towler, Rayleigh, Essex
??	F.3	<FF>. Stored RAF Stafford (ex-2156 ATC, Brierley Hill, West Midlands)
XP693	F.6	Barry Pover, Classic Jet Aircraft Company, Exeter Airport; potential flyer
XP694	F.3	Remains. Otterburn Training Area, Northumberland
XP701	F.3	<FF>. Robertsbridge Aviation Society, Mayfield, Sussex
XP702	F.3	Remains. Otterburn Training Area, Northumberland
XP703	F.3	Cockpit. Ground Procedures Trainer. LPG, Bruntingthorpe Airfield, Leics.
XP706	F.3	Lincolnshire Lightning Preservation Society, Strubby Airfield, Lincs.
XP708	F.3	Remains. DTEO, Pendine Ranges, nr Tenby, Wales
XP735	*F.3*	*Remains. DTEO, Pendine Ranges, nr Tenby, Wales (?Last reported 7.91)*
XP741	*F.3*	*Burnt at DFSCTE, RAF Manston, Ramsgate, Kent by 1994*
XP745	F.3	<FF>. Vanguard Engineering, Western Ave, Greenford, Middx.
XP748	F.3	Remains; no outer wings but fairly intact. DTEO, Pendine Ranges, nr Tenby, Wales
XR713	F.3	RAF Leuchars, Fife. 111 Squadron marks.

F.3 XR713 in 111 Sqn markings at Leuchars [MAP]

XR716	*F.3*	*Remains removed from RAF Cottesmore, Leics. 14.6.94*
XR718	F.3	Blyth Valley Aviation Collection, Walpole, Suffolk; under restoration (ex-BDRT)
XR724	F.6	Lightning Association, Binbrook Airfield, Lincs.
XR725	F.6	Charles Ross, Chestnut Farm, Binbrook village, Lincs.
XR726	F.6	<FF>. Privately owned, Harrogate, N. Yorks. *(Unconfirmed; last seen Rossington 3.93)*
XR727	*F.6*	*Reported scrapped at Wildenrath airfield, Germany*
XR728	F.6	Lightning Preservation Group, Bruntingthorpe Airfield, nr Lutterworth, Leics.
XR747	F.6	<FF>. Barry Pover/Ernie Marshall, Lightning Flying Club, c/o Plymouth Executive Aviation, City Airport, Plymouth, Devon
XR749	F.3	Ken Ward, Teesside Airport (ex-Middleton St George)
XR751	F.3	Roy Flood, Lower Tremar, St Cleer, nr Liskeard, Cornwall
XR753	F.6	XI(F) Squadron, RAF Leeming, N. Yorks
XR754	F.6	<FF>. Blyth Valley Aviation Collection, Walpole, Suffolk; for South Yorkshire Aircraft Museum, Firbeck
XR755	F.6	Ernie Marshall, Callington, Cornwall

F.6 XR755 at Callington in Ernie Marshall's garden [Graham Jackson]

XR757	F.6	<FF>. Trevor Garrod. Peaks Top Farm, New Waltham, nr Grimsby, S. Humberside
XR758	*F.6*	*Reported scrapped at RAF Laarbruch, Germany*
XR759	F.6	<FF>. Andrew Exton, Haxey, nr Gainsborough, Humberside
XR770	F.6	NATO Aircraft Museum, Peaks Top Farm, New Waltham, nr Grimsby, S. Humberside
XR771	F.6	Midland Aviation Museum, Coventry Airport. On loan from Magnatec Ltd.
XR773	F.6	Barry Pover, Classic Jet Aviation Company, Exeter Airport; potential flyer
XS416	T.5	NATO Aircraft Museum, Peaks Top Farm, New Waltham, nr Grimsby, S. Humberside
XS417	T.5	Newark Air Museum, Winthorpe, Notts.
XS420	T.5	Murray Flint. Fenland Aviation Museum, West Walton, nr Wisbech, Cambs.
XS421	T.5	<FF>. Nigel Towler, Rayleigh, Essex
XS422	T.5	Portsmouth Marine Salvage (ex-Haydon-Baillie Collection) *(For sale)*

T.5 XS422 taxies in after displaying at Boscombe Down Families Day, 21st September 1985
[Bob Finch]

XS451	T.5	Barry Pover, Classic Jet Aviation Co., City Airport, Plymouth, Devon; potential flyer
XS452	T.5	Mike Beachy Head, S. Africa. Purchased from Tony Hulls; potential flyer
XS456	T.5	T A Smith & Co., Wainfleet. Golf Range by Skegness to Boston Road

T.5 XS456 gleams in the sun after restoration at Wainfleet in 1994
[HT]

XS457	T.5	<FF>. Christian Brydges, Cleethorpes, S. Humberside
XS458	T.5	Tony Hulls, T.5 Projects, Cranfield Aerodrome, Beds.
XS459	T.5	Fenland Aviation Museum, West Walton, Wisbech, Cambs.
XS895	F.6	Remains dumped upside down at DTEO, Pendine Ranges, nr Tenby, Wales
XS897	F.6	South Yorkshire Aircraft Museum, Firbeck
XS898	F.6	<FF>. Tony Collins, Lavendon, Bucks. *(Remainder scrapped at Cranfield 1-2.12.94)*
XS899	F.6	<FF>. Charles Ross; on loan to Coltishall. *(Remainder scrapped at Cranfield 1.12.94)*
XS901	*F.6*	*Reported scrapped at RAF Bruggen, Germany.*
XS903	F.6	Yorkshire Air Museum, Elvington Airfield, nr York
XS904	F.6	Lightning Preservation Group, Bruntingthorpe Airfield, nr Lutterworth, Leics.
XS919	F.6	Ray Heming, Torpoint Shipping Co., Torpoint, Plymouth

F.6 XS901 at Bruggen in 1989—now scrapped. It carries the cautionary note on the intake-blank 'a/c tail heavy - without intake ballast'
[Peter Lawler]

XS922	F.6	Cockpit. Tony Dyer, Air Defence Collection, Amesbury, Wilts.
XS923	F.6	<FF>. Roy Jerman, Welshpool, Powys. *(Remainder scrapped at Cranfield 2.12.94)*
XS925	F.6	RAF Museum, Grahame Park Way, Hendon, N. London
XS928	F.6	BAe Warton Aerodrome, nr Preston. Stored (recently painted silver). Potential gate guard at Mill Lane entrance by Lightning Club
XS929	F.6	RAF Akrotiri, Cyprus. Gate guard
XS932	F.6	<FF>. Hugh Trevor, Lightning Preservation Group, Bruntingthorpe Airfield, Leics.
XS933	F.6	<FF>. Haydn Block, Terrington Aviation Collection, Terrington St Clement, nr Kings Lynn, Norfolk

F.6 XS935: a sad end for a once magnificent aircraft, looking undignified during scrapping at Rossington

XS936	F.6	Roy Flood. On pylon at Castle Air Grage, Liskeard, Cornwall

F.6 XS936 in pseudo XI/111 Squadron markings at Castle Air Garage, Liskeard
[Graham Jackson]

XV328	T.5	<FF>. Neville Martin, Bruntingthorpe, Leics. *(Remainder scrapped at Cranfield 5.12.94)*

All the Wildenrath Lightnings were reportedly scrapped upon the RAF leaving the airfield, and no BDRT F.2A's, T.4's or F.6's survive in Germany at RAF Bruggen and RAF Laarbruch (unless some were salvaged by the locals?).

Royal Saudi Air Force

54-650	T.54	Dhahran; on display
52-655	F.52	Technical Services Institute, Dhahran; on display
52-656	F.52	Stored for display, Riyadh
53-668	F.53	ZF577, Portsmouth Marine Salvage (ex-Haydon-Baillie collection)
53-670	F.53	ZF578, Removed by new owner upon closure of Wales Air Museum, Cardiff Airport
53-671	F.53	ZF579, Portsmouth Marine Salvage (ex-Haydon-Baillie collection)
53-672	F.53	ZF580, BAe Samlesbury Airfield. Gate guard

F.53 53-672 (ZF580) being placed on the gate at BAe Samlesbury
[BAe]

53-675	F.53	ZF581, Portsmouth Marine Salvage (ex-Haydon-Baillie collection)
53-676	F.53	ZF582, Portsmouth Marine Salvage (ex-Haydon-Baillie collection)
53-679	F.53	ZF590, Portsmouth Marine Salvage (ex-Haydon-Baillie collection)
53-681	F.53	ZF583, Solway Aviation Society, Carlisle Airport, Cumbria
53-682	F.53	ZF584, GEC-Marconi Ltd., South Gyle Works, Edinburgh. Gate guard

F.53 53-682 (ZF584) displayed outside the GEC Marconi (ex-Ferranti) works at East Gyle, Edinburgh
[HT]

53-683	F.53	ZF585, Portsmouth Marine Salvage (ex-Haydon-Baillie collection)
53-685	F.53	ZF591, Portsmouth Marine Salvage (ex-Haydon-Baillie collection)
53-686	F.53	ZF592, Portsmouth Marine Salvage (ex-Haydon-Baillie collection)
53-687	F.53	Khamis Mushayt; on display
53-688	F.53	ZF586, Portsmouth Marine Salvage (ex-Haydon-Baillie collection)
53-691	F.53	ZF587, Portsmouth Marine Salvage (ex-Haydon-Baillie collection)
53-692	F.53	ZF593, Robins AFB, Highway 247, nr Macon, Georgia, USA
53-693	F.53	ZF588, East Midlands Aeropark, Airport, Castle Donington
53-696	F.53	ZF594, North East Air Museum, Usworth, Sunderland, Tyne & Wear
53-699	F.53	Tabuk; on display
53-700	F.53	ZF589, Portsmouth Marine Salvage (ex-Haydon-Baillie collection) *[last Lightning built]*
55-711	T.55	ZF597, Portsmouth Marine Salvage (ex-Haydon-Baillie collection)
55-713	T.55	ZF598, Midland Air Museum, Coventry Airport
55-714	T.55	ZF595, Portsmouth Marine Salvage (ex-Haydon-Baillie collection)
55-715	T.55	ZF596, Portsmouth Marine Salvage (ex-Haydon-Baillie collection)

Note: The RSAF wish to set up a **Lightning Museum** at Khamis Mushayt and would like photographs of their pilots being trained at RAF Coltishall c.1968. Please contact: RE Sinclair, MODSAP, PNR3, Castlewood House, 77-91 New Oxford Street, London WC1A 1DS.

Royal Kuwaiti Air Force

Ten Kuwaiti machines were preserved as gate guardians and eight survived the Gulf War relatively intact.

55-410 T.55 'A' Badly damaged in the war and dumped at Ali Al Salem AFB

T.55 55-410 dumped at Ali Al Salem AFB in 1995
[Adrian Moitié]

55-411 T.55 'B' Preserved for Air Museum, Kuwait International Airport

T.55 55-411 sitting well down on her oleos on display at Kuwait International Airport on 1st April 1995
[Adrian Moitié]

53-412 F.53 'C' Displayed at Kuwait International Airport
53-415 F.53 'H'] 'H', 'J' & 'K' are displayed
53-416 F.53 'J'] on poles in a 'vic' formation
53-417 F.53 'K'] at Al Jaber AFB

F.53's of the Kuwaiti Air Force displayed in a spectacular 'vic' formation at Al Jaber AFB in 1995 (L-R 53-415 'H', 53-416 'J' and 53-417 'K')
[Adrian Moitié]

53-420 F.53 'N' Displayed at Kuwait Technology Institute
53-421 F.53 'O' Preserved for Air Museum, Kuwait International Airport
53-422 F.53 'P' Badly damaged at Messila Beach and dumped at Ali Al Salem AFB

F.53 53-422 'P' dumped at Ali Al Salem AFB, 1995
[Adrian Moitié]

53-423 F.53 'R' Gate guardian at Ali Al Salem AFB

There are rumours of a Kuwaiti Lightning in private hands; if so, the most likely candidate might be 53-418 'L' (G-AXEE), which was at Al Salem AFB and survived the war with little apparent damage.

Acknowledgements: Information for the above was gleaned from many sources, including the Lightning Association Review: Stewart Scott (Vol. 2 No.2 & Vol. 3 No.3); Graham Jackson (Vol. 1 No. 4); Mick Cameron (Vol. 2 No. 2); Peter & Margaret Thompson (Vol.1 No. 3); and from the invaluable 'Wrecks & Relics' (14th & 15th Eds.) by Ken Ellis (Midland Publishing Ltd). Karl Kaehler, Adrian Moitié, Tony Dyer and Roger Hurt also very generously supplied information.

LPG: INDIVIDUAL LIGHTNING HISTORIES

This information has been compiled from a variety of sources which are acknowledged below.
Some of the information is of a minor nature, e.g. dates of engine/air tests, APC's etc. but it is hoped that this may help enthusiasts, for example to date or identify an unknown photograph.

ABBREVIATIONS:

ADA = Unknown abbreviation on aircraft movement cards
ASF = Aircraft Servicing Flight
ASSF = Aircraft Storage & Support Flight, which (approx. 1983) became LESF
CWP = Contractor's Working Party
FEAF = Far East Air Force
LESF = Lightning Engineering Support Flight
MEAS = Mechanical Engineering Air Squadron (incorporated ASSF/LESF and ASF)
MOD(PE) = Ministry of Defence (Procurement Executive)
MU (Maintenance Units):
60 MU at RAF Leconfield
71 MU at RAF Bicester ("crash & smash" repairs, detachments visited Binbrook)
103 MU at Akrotiri

LIGHTNING HISTORIES

XR728 (English Electric Co. cockpit no. 95211)

17.03.65	f/f as F.3: Desmond 'Dizzy' de Villiers, Samlesbury to Warton for conversion to F.6
01.11.67	To: D 23 Sqn, Leuchars, delivered by Sqn Ldr Blackburn
05.09.68	D 23 Sqn, No. 1 engine compressor failure *(see Brian Carroll's article)*
17-23.05.69	D with 23 Sqn detachment via Gutersloh to Fassberg for Berlin corridor exercise Quicksand VIII
16.10.69	D with 23 Sqn to Akrotiri *(see Brian Carroll's article)*
31.03.70	D 23 Sqn at Akrotiri (used by XI Sqn detachment)
01.05.70	XI Sqn
08.06.70	D 23 Sqn
13.05.71	60 MU Leconfield, major servicing
24.11.71	To: D 56 Sqn, NEAF, Akrotiri; ferried out from Binbrook by 5 Sqn
22.08.73	103 MU Akrotiri
18.09.73	D 56 Sqn
31.01.75	D 56 Sqn transferred to 11 Group, Wattisham, UK
22.10.75	J 56 Sqn
18.11.75	J 56 Sqn - birdstrike on radar bullet
28.06.76	56 Sqn disbandment
01.07.76	Storage Flight, Binbrook
28.07.77	unmarked *[first camouflaged around this time?]*
14.10.77	Storage ASSF
27.06.78	5 Sqn
29.06.78	Birdstrike, port u/c
03.07.78	ASSF
12.07.78	J, no unit markings, air-test
01.08.78	D LTF - first F.6 for LTF and named 'Tiverton Trainer Target'
26.08.78	D LTF static at Binbrook Open Day
02.07.79	ASSF
06.10.81--05.11.81	Air-tests
12.11.81	Abingdon for repaint of grey/green camouflage
27.11.81	Binbrook MEAS
01.02.82	BA XI Sqn
03.06.82	5 Sqn
07.07.82	BA XI Sqn
28.08.82	BA XI Sqn in 9-ship at Open Day
07.03-13.04.83	BA XI Sqn to Akrotiri APC
03.08.83	MEAS
08.08.83	St Athan for repainting in dark sea grey
25.08.83	Returned Binbrook; MEAS
by 07.09.83	XI Sqn markings, uncoded
by 24.05.85	5 Sqn markings, uncoded
11.06.85	BF XI Sqn
29.07--04.09.85	BF XI Sqn APC, Akrotiri
By 19.12.85	BF XI Sqn stbd. nose roundel zapped XXX by 29 Sqn
22.4.86	MEAS
23.06.86--19.09.86	BF ASF Hangar for wing strengthening mods.
11.06.87	BS XI Sqn markings
16.06.87	BS XI Sqn markings, to engine test
? 06.87	BS XI Sqn markings, air-test
07.07.87	JS LTF markings, ASF hangar
16.07.87	JS engine test
21.07.87	JS air-test
04.08.87	JS allocated to 5 Sqn
22.08.87	JS led 9-ship at Last Last Lightning Show
7/8.09.87	JS at Valley MPC
29.09.87	JS in 5 Sqn 3-ship with XS932 AG and XR771 AN to mark final official Lightning F.3 flight
02.11.87	MEAS
03.11.87	JS to 5 Sqn

07.11.87	JS flew as #6 in 9-ship at 5 Sqn Reunion, replacing XR771 AN
09.11.87	JS to MEAS, ASF hangar for servicing
16.11.87	JS to LESF to complete servicing
01.12.87	JS air-test
03.12.87	JS flying with 5 Sqn
21.12.87	JS last to touch down after 5 Sqn final 4-ship
04.01.88	JS to XI Sqn
27.01.88	JS flown over Binbrook church at Neil MacLachlan's funeral by Furz Lloyd (last ever Lightning flight by a 5 Sqn pilot)
29.04.88	JS #4 in XI Sqn final 9-ship
28.05.88	JS in static at Beauvechain Air Show, Belgium
22.06.88	JS allocated to Lightning Preservation Group
24.06.88	JS delivered to Bruntingthorpe by Flt Lt Chris Berners-Price

Total Hours: 3708.40

XR728 'D' and XS932 'J' of 56 Squadron on approach with overwing tanks fitted.
[Terry Senior]

Rare photograph of XR728 in 1978 when with the Lightning Training Flight and marked up with the title "Tiverton Trainer Target" on a blue square.
[Ian Black]

XR728—Incidents

18.11.75	J 56 Sqn: Bird-strike on radar bullet at night. Bird and metallic debris ingested by both engines. 130 man-hours to repair.
12.01.76	J 56 Sqn: Flash-fire in flight during fuel transfer, Fire 1caption illuminated. Caused by leaking pipe-seal in Fueldraulic system. Fire drill on No.1 engine actioned, only one fire-bottle worked.
29.06.78	5 Sqn: Bird-strike in port undercarriage area during landing, slight damage.
03.12.81	MEAS Air test?: FCS Computer tripped out causing aircraft to barrel-roll and dive in 'attitude' mode.
15.12.81	MEAS Air test?: Air Turbine Gearbox seized, main bearing failed due to out-of-balance condition.
19.08.82	BA XI Sqn: Undemanded rolling and pitching. Pilot switched off autopilot and autostabiliser and regained stable flight. 4-channel actuator amplifier subsequently replaced.
26.01.83	BA XI Sqn: Probe tip broke off during air-air refuelling. There had been several probe engagements the previous day in turbulent conditions. *(Ref. Mike Hale's 'Scramble!' article!)*
26.05.83	BA XI Sqn: At 400 kts, 500 ft a bird glanced off the fuselage below the canopy. No damage.
11.02.86	BF XI Sqn: During undercarriage retraction, starboard red light stayed on. Thought to be due to damp in wiring after aircraft had stood for 8 days in unheated QRA hangar.

Note: At some point, possibly when coded BF of XI Sqn (or just before), it is thought the canopy was lost and struck the fin. A large repair to the port side of the fin is still visible today.

(Right) XR728 'J' taxies out for air test in July 1978. (Below) Composite photograph shows XR728 gleaming in a brand new dark sea grey paint scheme in 1983 - that ventral tank would never look as clean again! The T.5 behind is XS452.
[Both Terry Senior]

(Right) XR728 'BF' with nose roundel zapped with the triple-X of 29 Sqn, 19th December 1985 [MAP]
(Below) XR728 was briefly coded 'BS' in June 1987 before becoming 'JS' [Trevor Garrod]
5 Sqn's final 9-ship above the fog-bound Lincolnshire Wolds, 7th November 1987, with 'JS' (nearest) a late substitute for a u/s aircraft [Geoff Lee, BAe}

XS904 (English Electric Co. cockpit no. 95250)

26.08.66	f/f Desmond de Villiers, Samlesbury to Warton for storage
19.04.67	Coded A, XI Sqn, Leuchars
Date?	A XI Sqn detached to FEAF
05.02.69	A XI Sqn returned from detachment to FEAF
12.08.70	60 MU Leconfield, major s ervicing
21.12.70	A XI Sqn
21.05.71	71 MU
05.07.71	A XI Sqn
(22.03.72-22.07.72	XI Sqn transfer from Leuchars to Binbrook)
16.10.72	Loaned to MOD(PE)
26.10.72	A XI Sqn
19.08.74	60 MU
09.01.75	A XI Sqn
early 75	Akrotiri with XI Sqn re Cyprus war to replace 56 Sqn on QRA
27.04.76	Storage Flight, Binbrook
[first camouflaged around this time?]	
29.07.76	A? XI Sqn
30.07.76	Storage Flight
04.08.76	A? XI Sqn
by 07.77	B XI Sqn
07.04.78	ASSF Store
01.02.80	D XI Sqn
14.03.80	D XI Sqn QRA to Faroes Gap, 2 x Bear F *(see Pete Hodgson article)*
15-18.07.80	D XI Sqn Akrotiri APC
30.07.80	5 Sqn
05.09.80	D XI Sqn
17.08.81	MEAS
04.12.81	ADA
10.05.83	MEAS; LTF hangar, no markings
21.08.84	XI Sqn
05.09.84	BM XI Sqn
02.01.85	Skidded off taxiway on ice, shearing off nosewheel oleo
by 02.85	back in service
29.07-04.09.85	BM XI Sqn to Akrotiri APC
13.05.86	MEAS
28.05.86	5 Sqn
09.06.86	MEAS
14.07.86-14.10.86	Wing strengthening mods.
06.01.87	ASF hangar, ex-XI Sqn uncoded
18.02.87	ASF Air-test
20.02.87	XI Sqn
16.03.87	BQ XI Sqn (first Lightning to carry this code)
03-07.08.87	BQ XI Sqn exchange to Reims, FAF EC.30 Mirage F.1C
22.08.87	BQ XI Sqn took part in 9-ship at Last Last Lightning Show
02-10.09.87	BQ XI Sqn exchange with 323 Sqn RNethAF at Leeuwarden
21.10.87	ASF servicing, Check 2A
14.12.87	Overwing tanks fitted, ASF
15.12.87	First flight with o/w tanks, then to LESF and No.1 hangar on BAe test work with Tornado ADV Foxhunter AI.24 radar
01.02.88	LESF hangar: Check 2A, Chief Technician Ian Donaldson's team
08.04.88	Delivered to BAe Warton (unmarked) by Flt Lt Dick Heath for continuation of radar trials (3 days before XR724)
12.04.88	Assigned MOD(PE) Warton; cumulative flying time 3959.11 hr, 3881 flights
18.04.88	First target flight with Tornado A.03, pilot Peter Gordon-Johnson
24.08.88	Achieved 4000 flying hours on 3912th flight, flown by Peter Gordon-Johnson
11.10.88	Hook came down at 400kt, pilot Colin Rae
06.01.89	Parking brake jumped out of runner during t/o run. Derek Reeh lost directional control until lever returned to runner; flight aborted
08.05.91	Derek Reeh took the RHAG after the brake chute candled
11.10.91	Flown to RAF Wildenrath by Derek Reeh; fin zapped by 19 Sqn
14.10.91	Returned to Warton by Peter Gordon-Johnson
31.01.92	4000th flight (4121.30 hr); target A.03
13-14.06.92	Boscombe Down Air Tournament International, static
16.12.92	Last Lightning 3-ship with XP693 and XR773, Warton (fly-throughs at RAF Leeming, BAe Samlesbury & Blackpool Squires Gate)
22.12.92	Last static line-up of the 4 Warton Lightnings
21.01.93	Delivered to Lightning Preservation Group by Peter Orme

Hours: 4161.51 (4034 flights); 203 hr while based at Warton

(Right) XS904 'A' of XI Sqn in 1975 (with XS933 behind) [MAP]
(Below) After camouflage was introduced, possibly 1976 [MAP]
(Bottom) Superb shot of XS904 'BQ' and XS898 breaking after take-off from Binbrook in 1987. XS898 became part of "Arnold Glass's Air Force" but was scrapped at Cranfield in December 1994; however the cockpit survives with Tony Collins at Lavendon.
[Dave Drummond]

XP703 (English Electric Co. cockpit no. 95157)

28.09.63	f/f Desmond de Villiers, Samlesbury to Warton
08.06.64	Coded G, 74 Sqn Coltishall
14.12.65	'Disposal account'
15.06.66	60 MU (Mod. 4052)
28.06.66	G 74 Sqn
By 10.10.66	G 74 Sqn, No. 2 Hangar, Wattisham for BAC Modification Team
14.11.66	G 74 Sqn marks, awaiting air test
22.11.66	S 56 Sqn
15.12.66	'Disposal account'
09.02.67	S 56 Sqn
11.04.67	With 56 Sqn to NEAF, Akrotiri
01.11.67	60 MU Leconfield
04.07.68	Wattisham to 56 Sqn NEAF, Akrotiri
09.68	Recoded S 56 Sqn
12.09.69	60 MU
10.01.70	56 Sqn
06.09.71	60 MU servicing and storage
11.08.72	G 29 Sqn, Wattisham
14.05.73	'Disposal account'
16.08.73	71 MU
02.10.73	G 29 Sqn
15.11.74	29 Sqn disbands
01.01.75	Storage flight, Wattisham
04.02.75	60 MU
23.05.75	MOD (PE). Flown by Sqn Ldr R.W. Turbin to BAC Warton for structural tests, having completed 3000 flying hours. Retained for structural testing to the equivalent of 3500 h.
11.75	Dumped outside at Warton
06.76	Fuselage in 2 sections
01.09.79	Cockpit (29 Sqn markings:'Flt Lt PSG Adams') on trolley with remains of fuselage and wings nearby.
Date?	Conversion to computer-controlled Engine Ground Testing Simulator for Saudi Support Unit
23.08.89	Lightning Preservation Group, Bruntingthorpe

(Above) XP703 as 'G' of 74 Squadron [APN]
(Left) Engine Ground Testing Simulator as used by the BAC Saudi Support Unit. 'A' and 'B' type Avpin starter failures (known as 'whee' & 'whee-phut'!) can be input to the computer control unit by the instructor via the rear panel (just visible). Engine noise at various revs is produced by the computer changing the speed of playback of engine noise on a continuous loop tape cassette. [HT]

XS932 (English Electric Co. cockpit no. 95265)

09.04.67	f/f Roland Beamont, Samlesbury to Warton
22.06.67	Coded H of XI Sqn, Leuchars
25.07.69	CWP
01.10.69	H XI Sqn
17.11.70	103 MU, Akrotiri
19.02.71	H XI Sqn
14.05.71	60 MU Leconfield, major servicing
25.10.71	J 56 Sqn, NEAF, Akrotiri
14.10.72	103 MU
31.10.72	J 56 Sqn, Akrotiri
31.01.75	J 56 Sqn transfer to 11 Group, Wattisham, UK
25.02.76	R 56 Sqn, Wattisham
01.07.76	Storage Flight, Binbrook
[First camouflaged around this time?]	
27.09.76	ADA
19.04.77	ASSF storage
by 08.77	Coningsby, stored BBMF hangar
27.02.79	XI Sqn, Binbrook
07.03.79	ASSF
13.03.79	5 Sqn
23.04.79	F 5 Sqn
30.01.80	F, ASF hangar, stored for ASSF
04.08.83	No markings, air test
25.08.83	To St Athan for dark sea grey repaint
12.09.83	Returned Binbrook
19.09.83	MEAS
20.09.83	Uncoded, air test
21.09.83	XI Sqn
29.09.83	AM 5 Sqn
by 14.10.83	AC 5 Sqn
by 19.10.83	AC 5 Sqn fitted with overwing tanks
02.12.83	AC: Sqn Ldr Dave Carden's 3000th hr on Lightnings
02-03.02.84	AC 5 Sqn MPC Valley
11.03.85	MEAS
17.05.85	LTF hangar for wing strengthening mods (carried out 23.05.85-14.08.85)
19.12.85	Engine test
20/23.12.85	Engine tests with 5 Sqn markings
10.01.86	AG 5 Sqn
31.05.86	AG 5 Sqn static at Bournemouth Hurn Air Show
11.10.86	AG 5 Sqn in 9-ship at 21st Anniversary Reunion of 5 Sqn on Lightnings (pilot: 'Peg' Clowes)
09-24.02.87	AG 5 Sqn APC Akrotiri
05.87	AG 5 Sqn exchange with 350 Sqn, BelgAF Beauvechain
22.06-31.07.87	AG 5 Sqn APC Akrotiri
22.08.87	AG 5 Sqn in Hangar 2 during Binbrook Open Day
29.09.87	AG in 5 Sqn 3-ship with XR728 JS and XR771 AN to mark final official Lightning F.3 flight
02.11.87	MEAS
07.11.87	AG Last flight in 9-ship at 5 Sqn Final Lightning Reunion
Hours: 3,655.30	
09.11.87	LESF hangar for spares recovery and storage
by 11.05.88	Decoy line (markings painted out)
17.05.88	'Scrap'
24.06.88	Allocated to Tanks and Vessels Industries, Ltd.
14.07.88	Transferred to TVI Ltd., Rossington nr Doncaster
09.11.91	Cockpit section to Bruntingthorpe

Acknowledgements

These histories were compiled from various sources, to whom thanks are due: Aircraft Movement Cards, copies of which were kindly provided by Mr G. Day at the Air Historical Branch (RAF) and Mr Mungo Chapman at the Royal Air Force Museum, Hendon; 'Lightning' by Roger Lindsay (pub. Ian Allan); 'Humberside Air Review' published by the Humberside Aviation Group; 'Roundel' magazine published by the British Aviation Research Group; and 'Lightning Log' by Alan Williams in the Lightning Association's Review Vol. 3 No. 2.

XS932 spent its first four years with XI Squadron at Leuchars coded 'H'. [MAP]

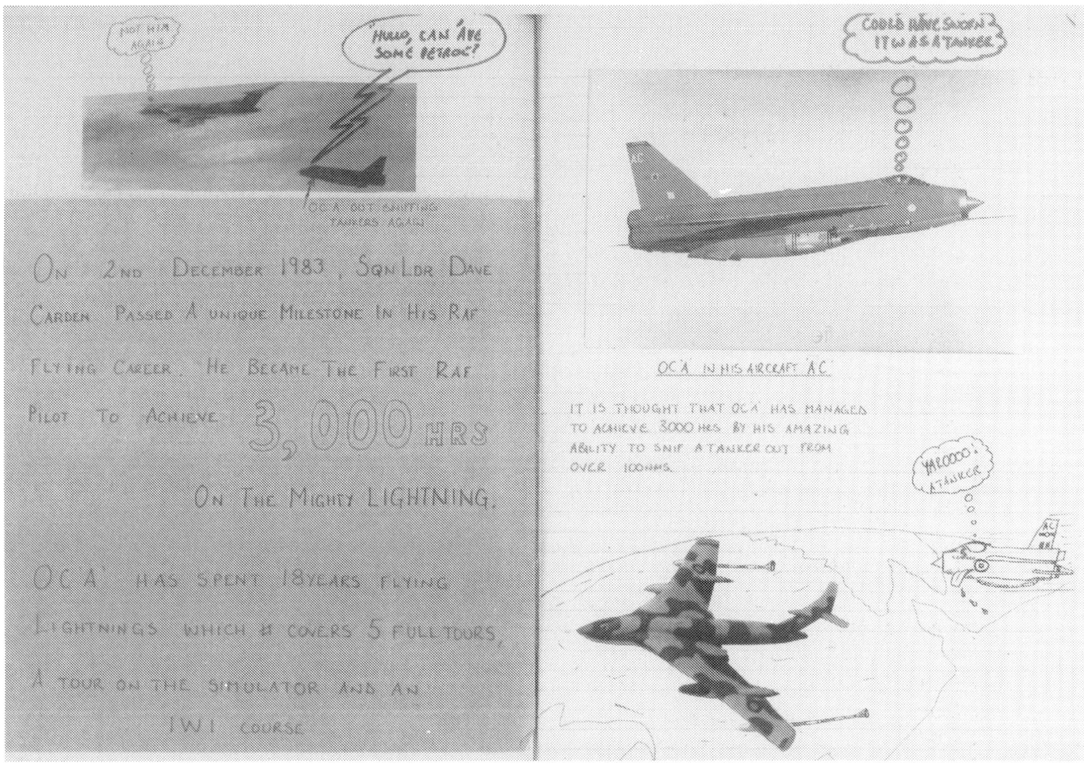

(Above) The 5 Squadron diaries commemorated Sqn Ldr Dave Carden becoming the first RAF pilot to reach 3000 Lightning hours on 2nd December 1983 in XS932 'AC', 'his' aircraft. [courtesy Flt Lt Mark Simmonds/5 Squadron] *(Below) XS932 'AG' during its final flight, 7th November 1987, accompanied by XS899 'AF' and XR727 'AB'.* [Geoff Lee, BAe] *(Bottom left) Ken Ward lends a hand with cockpit recovery at Rossington in 1991 (XR726 & XS935 behind).* [HT] *(Bottom right) The cockpit of XS932 at Bruntingthorpe.* [HT]

BIBLIOGRAPHY

Air Extra No. 14. *The BAC Lightning.* Ian Allan. ISSN 0307-7411.
Allan, Chris. *Fast Jets.* Osprey Publishing, 1986. ISBN 0-85045-661-4.
Allan, Chris. *Fast Jets 2.* Osprey Publishing, 1987. ISBN 0-85045-754-8.
Allan, Chris. *The Royal Air Force in Colour.* Ian Allan, 1988. ISBN 0-7110-1725-5.
Beamont, Roland P. *Testing Years.* Ian Allan, 1980. ISBN 0-7110-10721-2.
Beamont, Roland P. *EE P1 Lightning.* Ian Allan, 1985. ISBN 0-7110-1471-X.
Beamont, Roland P. *Fighter Test Pilot.* Patrick Stephens Ltd., 1986. ISBN 0-85059-850-8.
Beamont, Roland P. *Testing Early Jets.* Airlife Publishing, 1990. ISBN 1-85310-158-3.
Beattie, John. *Drama in the Air.* Robson Books, 1989. ISBN 0-86051-664-4.
Black, Ian. *Lightning.* Airlife Publishing, 1988. ISBN 1-85310-045-5.
Black, Ian. *Jet Combat.* Osprey Publishing, 1988. ISBN 0-85045-839-0.
Black, Ian. *Last of the Lightnings.* Haynes Publishing, 1996 (in press).
Brookes, Andrew. *Crash!* Ian Allan, 1991. ISBN 0-7110-1965-7.
Caruana, Richard J. Close-up Classics: 2, *The BAC Lightning.* Modelaid International Publications, Malta. ISBN 1-871767-032.
Chesneau, R. & Rimell, R. Aeroguide No. 8: *BAC Lightning FMk3/Mk6.* Linewrights Ltd., 1984. ISBN 0-946958-07-6
Darling, Kev. *Lightning, The Operational History.* Airlife Publishing, 1995. ISBN 1-85310-521-X.
Ellis, Ken. *Wrecks & Relics.* Midland Publishing, 1994. ISBN 1 85780 025 7.
Famous Airplanes of the World No. 75: *BAC Lightning.* Koku Fan Special, July 1976.
Gardner, Charles. *British Aircraft Corporation.* B.T. Batsford Ltd., London, 1981. ISBN 0-7134-3815-0.
Gunston, Bill. *Early Supersonic Fighters of the West.* Ian Allan , 1976. ISBN 0-7110-0636-9-103/74.
Halpenny, Bruce Barrymore. *EE/BAC Lightning.* Osprey Air Combat, 1984. ISBN 0-85045-562-6.
Jackson, Paul A. Aviation News Special: *BAC Lightning Marks 1-6.* Alan W. Hall Publications, 1977. ISBN 0-89402-035-8.
James, H.G. Profile No. 144: *The English Electric P.1 and Lightning 1.* Profile Publications.
Levy, R. & Ward, R. Aircam Aviation Series No. 37: *English Electric (BAC) Lightning.* Osprey Publications Ltd., 1972. ISBN 85045-138-8.
Lindsay, Roger. Aviation News Mini Monograph: *English Electric Lightning FMk1/1a.* Alan W. Hall Publications.
Lindsay, Roger. *25 Years of Lightnings.* In: Aircraft Illustrated Annual 1980, ed. M. Horseman. Ian Allan. ISBN 0-7110-0936-8.
Lindsay, Roger. Aircraft Illustrated Special: *Lightning.* Ian Allan, 1989. ISBN 0-7110-1825-1.
Moulds, S/L Gordon. *Lightning Conversion Units 1960-1987.* Prospect Litho Ltd, 1993.
Philpott, Bryan. *EE/BAC Lightning.* Patrick Stephens Ltd., 1984. ISBN 0-85059-687-4.
Ransom, Stephen & Fairclough, Robert. *English Electric Aircraft and their Predecessors.* Putnam, 1987. ISBN 0-85177-806-2.
Reed, Arthur. Modern Combat Aircraft: 5, *BAC Lightning.* Ian Allan, 1980. ISBN 0-7110-0988-0.
Reed, Arthur. RAF Aircraft Today: 2, *Lightning.* Ian Allan, 1984. ISBN 0-7110-1407-8.
Rentoul, Ian & Wakeford, Tom. *Final Days for the Lightning.* In: Aircraft Illustrated Military Aviation Review, 1988. Ian Allan. ISBN 0-7110-1801-4.
Scott, Stewart. *Lightning Photo Album, Vol. 1.* GMS Enterprises, Peterborough, 1994.
Scott, Stewart, & Jackson, John. *Offence to Defence!* The History of RAF Binbrook. GMS Enterprises, Peterborough, 1990. ISBN 1-870384-07-5.

XS928, formerly used in the Tornado ADV Foxhunter radar trials, was repainted in 5 Squadron markings in 1996 prior t[o] becoming the gate guardian at BAe Warton [British Aerospace/North West Heritage Group]

Index of Main Text

Note: This Index does not cover the 'Lightning Survivors' and 'Acknowledgements' sections.

(Above) Let's Go! Mike Hale eager for the off [John Dunnell via Mike Hale]
(Below) ZF593 (53-692) in 29 Sqn markings at Robins AFB Museum, Georgia in 1996 [MAP]

Index of Lightning Airframes referred to in Main Text

See also 'Lightning Survivors' section.

Nice shot of F.3 XP694....but red faces for the aircraft painters when they examined that serial under the port wing!

[British Aerospace/North West Heritage Group]

(Top) XI Squadron F.6's approaching the coast near Leuchars; XS918 'B' leading, XS932 'H' to right, XS904 'A' to left and XS930 'F' in the No. 4 position [Crown Copyright/MOD]

(Bottom) XS904 at low level and high speed over Binbrook Airfield in 1985 [John Dunnell via Mike Hale]

(Top) F.53 on trials with 1000lb bombs and here seen with rocket pack doors open. G27-56 was registered G-AWON for Farnborough 1968 and later became 53-686 and ZF592; it is now with Portsmouth Marine Salvage. (Below) A line of Kuwaiti and Saudi F.53's at Warton before delivery; F.53K 53-416 nearest camera is now preserved at Al Jaber AFB. [both BAe/North West Heritage Group] *(Bottom) XS904 and Keith Hartley on 23.9.92. With the recent more positive approach from the CAA and recalculation of 904's fatigue life....who knows—we just **may** see this sight again!* [Andy Rickard]